The Sublime Engine

The
SUBLIME
ENGINE

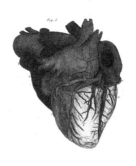

A Biography of the
HUMAN HEART

STEPHEN AMIDON
AND THOMAS AMIDON, MD

RODALE

Book design by Sara Stemen

Library of Congress Cataloging-in-Publication Data

Amidon, Stephen.
   The sublime engine : a biography of the human heart / Stephen Amidon and Thomas Amidon.
      p.   cm.
   Includes bibliographical references and index.
   ISBN 978-1-60529-584-8   hardcover
   1. Heart—Popular works.   2. Heart in art.   3. Heart in literature.   4. Heart—Religious aspects.   I. Amidon, Thomas W. II. Title.
   QP111.4.A45   2011
   612.1'7—dc22   2010030227

Distributed to the trade by Macmillan

2   4   6   8   10   9   7   5   3   1   hardcover

We inspire and enable people to improve their lives and the world around them.

To our parents, Bill and Bess

Contents

Introduction

Imagine that you are a primitive. You have just clawed your way to the top rung of the evolutionary ladder. You have achieved humanity. Your mind is developing a capacity for the self-consciousness that will be your defining characteristic as a member of the species *Homo sapiens.* You are aware of yourself as being distinct from others, with private thoughts and complex relationships with those around you. You have a self.

You lie in your cave after a long day's hunting. Although you do not yet possess anything that could be called language, you find yourself able to string together connected thoughts. As night comes and sleep approaches, you use this new awareness to take stock of your body, perhaps probing a bruised limb or savoring the taste of recent food. And then, in the dark silence, a terrible and awe-inspiring realization strikes you. It is something you have never before noticed in your grueling daily quest for simple survival.

There is something alive inside you.

Right there, pulsing beneath your ribs. It hammers out a steady rhythm that reverberates through your musculature. If you touch your neck or your wrist or your leg, you can feel its report. It is as if a small but powerful beast has lodged inside your chest. The realization excites and frightens you. With this surge of emotion, the beast responds by quickening its activity—as if it senses your fear.

For the next few days, you go about your routine. You have no choice. You must, after all, survive. But it is always there, as you hunt and feast and rest, this churning presence inside your ribs. Speeding up sometimes, slowing down in other situations—but never resting, not for an instant. In the middle of the night, when everything else is quiet, it can be as loud as thunder.

And then, in a moment as terrible and wonderful as when you first became aware of this pulsing creature, you understand something else. This is no foreign body. This is no predatory stranger trapped within your bones. It is part of you. It *is* you. The center of that individuality you have only recently understood. You have discovered yourself. You have discovered your heart.

FOR as long as we have been self-aware, we have been in awe of the fact that there is something so vital, so alive, within our bodies: a relentlessly active core with a will of its own. An animating essence that does not obey our commands the way our hands do, or our eyelids, or even our lungs. A link to the universal motion surrounding us, the tides and stars and winds, with their puzzling rhythms and unseen sources. Once this awareness dawned, it would have been impossible for us ever again to look at ourselves or the world the same way.

The heart is a mystery and a miracle. It beats roughly every second of our lives—two and a half billion times during an average lifespan. It does not rest, pumping around seventy-four gallons of blood every hour once we become adults. Although it weighs only about fifteen ounces, it is immensely strong—the amount of energy it generates in a single day could drive a car twenty miles.

And it can stop at any moment. It can stop when we are newborn babies asleep in the crib or when we are healthy teenagers walking off the practice field on a sultry August afternoon. It can arrest when we are shoveling snow in our middle age or when we are a hundred years old and our minds have been swept of memory for a decade. The heart is the ultimate arbiter of our lives. When it calls time, the game is over.

What follows is a biography of this remarkable machine. Although one of the authors is a cardiologist, it is not a medical history, despite the fact that science plays a key role in the story. And while the other author is a novelist and critic, it is not a cultural study, though works of literature, art, and film are also significant elements in the book. Nor is it fiction, though some of the stories told here are so dramatic that they will be presented in narrative form in order to fully capture the power they can wield over our imagination.

Rather, this book is an account of the different ways that we have thought about the heart ever since it first took root in the Western imagination. Whether it causes a scientist to investigate the origins of the pulse, or inspires a poet to wonder what makes that pulse race *faster* when he sets eyes on his beloved, the cardiac muscle has always been the object of passionate inquiry. *The Sublime Engine* is the history of that passion. The authors have, in the interest of focus and space, confined this biography to the Western tradition, although there is a rich tradition of interest in the cardiac organ in other cultures, from the elaborate pulse lore of traditional Chinese medicine to the Nadi Vijnana pulse science of the Indian subcontinent; from Babylonian creation myths to the bloodthirsty heart sacrifices of the Aztecs.

Stories lie at the center of this biography. The heart is filled with more than just the blood it pumps. It is also the source of a vast supply of stories. They flow through it; they echo around its chambers. Some of these stories are historical, capturing the people whose insights and experiments have furthered our understanding of the cardiac muscle. Pioneers like Hippocrates, who turned his back on superstition and magic to create the practice of medicine, which would eventually allow us to know and even control the organ beating in our chests. Or Werner Forssmann, the young German intern who defied received wisdom by inserting a catheter into his heart at a time when most of his colleagues thought such a procedure would cause certain death. Or the unassuming Minnesota surgeon C. Walton Lillehei, who stitched the circulatory systems of his young patients to those of their parents—effectively using the latter as heart-lung machines—in a bold attempt to save the lives of doomed children. Or René Laënnec, the musically trained Parisian physician who first heard the symphony of heart sounds by rolling up a stack of paper and placing it against a shy female patient's chest. And then there is the greatest genius of them all, Sir William Harvey, the personal physician to two kings, who forever changed the way we think about the heart by simply placing a finger on his forearm. By telling the tales of these adventurers, we can begin to illustrate the vastness and wonder of the territory they explored.

Some of the heart's stories are anatomical. Because the living cardiac muscle is always in motion, its every beat is a kind of story, with a beginning, a middle, and an end. Nature provides nothing more dramatic than the motion of the heart. Everything about it has the dynamism of a good narrative. The way a red blood cell can be propelled through the circulatory system in about the time it takes an Olympic athlete to run 200 meters. Or how white blood cells will aggressively attack plaque in a coronary artery, shutting down the blood supply to the heart muscle and leading to that most terrifying of bodily events—the heart attack. And then there is the way the heartbeat itself is formed, a delicate dance of electrically charged sodium, potassium, and calcium ions as they pass back and forth through specialized channels in the heart cell's membrane. Or how, after years of reliable thumping, that steady beat can, without warning, dissolve into a riot of utter chaos—fibrillation—that can kill the heart's owner in a matter of minutes.

There is yet another set of stories that illuminate the heart. These come not from the laboratory, operating theater, or examination room but, rather, from the deepest precincts of the human imagination. Since antiquity, the heart has been the artist's central image for describing those qualities that make us most human. It has played an equally rich role in the work of mystics, preachers, and saints as they seek to describe our relationship with the eternal. The heart is more than just an ingenious pump— it is a metaphor of almost limitless profundity. It is an engine not just of blood but also of sublimity. We use it to signify our deepest thoughts and emotions when words alone do not suffice. The ancient Egyptians saw the heart as the part of a human being that ascended to heaven, carried by a winged beetle; the more earthly minded Greeks saw it as the home of a man's courage and loyalty.

Believers in Yahweh and Jesus Christ deemed it the organ by which a person communicates with his or her Creator; poets saw it as the means of forming emotional bonds with other human beings. Dramatists, meanwhile, showed an odd propensity for depicting it as the source of our darkest, most unruly passions—a fountain of chaos and tragedy.

These days, in the post-Freudian era of the mind, there is a growing school of thought that tells us we should know better than to endow the heart with these elusive qualities. It is a pump, plain and simple; a muscle, utterly transparent to imaging technology and explicable in the language of biochemistry and physiology. Its central symbolic function nowadays is to represent our quest for bodily health. To this way of thinking, the heart has more meaning on a cereal box than in a painting; it has a bigger role in a television commercial than in a prize-winning play. Its central mystery is not what sort of spiritual truth it contains but how we can keep it beating for as long as possible.

And yet our belief in the heart's transcendence endures. We continue to use it as a metaphor for love and courage and devotion without the slightest risk of confusion. A girl scratches its shape on a boy's notebook with a pen; a basketball player taps the left side of his chest with his fist after making a clutch shot—there is no confusion about what is being communicated. Four thousand years of usage have lodged the heart at the core of our imaginations. It is not likely that it will be evicted anytime soon.

CHAPTER I

Ancient Heart

The Island of Kos, Greece, 399 BC

He does not want to die. He is not ready. He still wants to breathe the sea air as he arrives at the port each morning and drink wine after a long day's work. He wants to sleep with his wife for many more nights and watch his son become a man.

But Nikias can feel death coming. He cannot walk more than a short distance without needing to stop to catch his breath. On some days it feels as if a massive slab of marble has been laid across his ribs. There is also pain in his arm and neck now. And then, two days ago, while helping to unload a shipment of ivory, he blacked out. If his son had not been there to catch him, he would have plunged right into the Aegean.

It makes him feel *old*. Granted, he is thirty-nine, and his father and grandfather were dead by this age, both felled by lightning in their chests. But he has always believed that he would outlive them by many more years than this. He will never forget the moment his father died—the way that powerful man gasped for air as he tore at his chest; the panic in his eyes, replaced at the end by a look of terrible emptiness. And then he was gone forever.

After his own collapse, Nikias knew he needed to get help. At his wife's urging, he went to the local doctor. But the old fool had said there was nothing to be done, that his fate was in the hands of the gods. There are other healers nearby, root cutters and magicians and soothsayers, but Nikias has even less faith in them.

Which leaves the temple of Asklepios, the healing god. People speak of the miracles that happen there. But who has time to visit a temple? Nikias's business might be thriving, but that is only because he is always there. Every day seems to bring a new crisis. The pressure on him is immense. And yet something must be done, or else he will soon be joining his father and his grandfather. In the quietest hour of the night, when everyone else is asleep, he can feel the blackness creeping over him.

So he decides to go. The temple is close enough to walk to under normal circumstances, but his weakness means that he must travel by oxcart. His son takes him. It is midday when the temple comes into view, situated on a cypress-strewn hill overlooking the sea. He has seen it before, but only in glimpses. It is larger than he recalls. Its marble glistens majestically in the spring sunlight. His son stops the cart at the bottom of the steep steps leading up to the temple's high wall. A year ago Nikias would have bolted up this hill, but now he must grip the boy's strong arm. By the time he nears the top, he is gasping for air. Sweat coats his skin; his chest has never felt tighter. He feels death within him, spreading through his muscles and bones.

After they knock on the heavy wooden door, Nikias notices a ramshackle encampment that abuts the wall. It is populated by beggars and cripples. A few hobble about on crutches; most sprawl on filthy blankets. A corpse, partially covered with sticks and hay, lies in the nearby weeds. A haggard woman cries over it, switching a cypress branch in a futile attempt to keep flies away.

The breeze shifts and Nikias can smell the stench. He cannot tear his eyes away from the body.

And then he realizes that a man is watching him from the encampment. He appears to be Nikias's age. He is different from the others. His clothes are clean; his beard is neatly trimmed. He is in good health; he is certainly no beggar. There is confidence and intelligence in his eyes. An oddly shaped bronze cup hangs from his belt, along with other instruments whose purpose Nikias cannot fathom. The man's scrutiny of Nikias is intense; he looks as if he has something urgent to say. Nikias is tempted to speak with him, but just then the door opens.

It is a goatish old man dressed in a *himation*, the long embroidered cloak of a senior priest, who stands before him. Still a little breathless, Nikias tells him why he has come. The priest nods sympathetically, then tells Nikias what offering will be required to gain entry. The price is steeper than those steps. But he needs to be cured, so he pays.

After telling Nikias's son to return in two days, the priest ushers the merchant through the gate. The temple grounds cover several terraced acres. As he leads Nikias along a shaded path toward the main temple, the priest describes what will happen. First Nikias will be cleansed with holy water. Then the two of them will make a sacrifice to Asklepios. After this they will pray to the god himself. Finally, Nikias will be led to the place where he will spend the night. If he is blessed, the god will speak to him in his sleep.

Nikias is tempted to ask when he will actually be *cured*, but he decides to remain quiet for now. They draw near the temple. Its colonnade is made of the highest-quality marble. Healing must be a good business, the merchant thinks. There are also numerous outbuildings. Animals, priests, and visitors gather on the temple steps or wander the shaded pathways. Some of the visitors appear to be seriously ill, though every one of them is attended by priests or slaves. There are no dead bodies in the weeds here.

The priest and Nikias stop first at a small wooden hut that contains a sacred well. A man, another visitor, emerges just as they arrive. He is dressed in a brilliant white robe. His wet hair glistens in the sun; his expression is serene. Nikias is heartened—perhaps there is something to this temple business after all. The priest chants a ritual prayer as they step inside. Flickering lanterns light the hut. The smell of incense is overpowering. Two slaves greet them. As one draws well water, the second takes Nikias's clothes.

That is when he sees the snakes. The room is full of them—writhing, coiled, tongues flicking lazily. The merchant's chest starts to hammer; his throat tightens as if a great fist is clenching it. He hates snakes. He has seen what happens when they strike from amid cargo. He backs toward the door, but the old priest stops him. There is nothing to fear, he claims; the snakes are harmless. They not only keep demons out of the water, their bites can also cure. Nikias agrees to stay, though he secretly resolves that he will leap into that well before he lets himself be bitten.

After Nikias is washed and then dressed in a pure white robe, the priest leads him to a massive stone plinth just outside the temple's main entrance. The altar is adorned with friezes of the gods and inlaid with gold medallions. Bleached animal skulls are scattered around it. Other priests and slaves await them. The priests chant incantations as the slaves lift a bound, writhing goat onto a bloodstained slab beside the altar's smoldering fire. Moving with a butcher's skill, one of the priests slits the creature's throat with a large knife, then expertly dismembers the quivering beast, tossing its thighs and flesh onto the altar fire. They crackle and hiss. The priests whisper among themselves as they examine the steaming entrails.

No one bothers to tell Nikias what has been divined. He is led to the temple, where he is confronted by the great ivory figure of Asklepios staring down from a majestic golden throne. The god is powerfully built, with broad shoulders and a mighty beard. In one hand he grips a staff entwined with snakes. The other rests on the head of a golden dog. Terra-cotta objects surround him, each depicting a body part. Nikias sees an ear, a hand, an eye, a penis. The old priest explains that these are gifts from former penitents, commemorating the healed parts of their bodies. He then commands Nikias to touch the statue to receive the god's healing power. The ivory ankle is cold and lifeless. And the pain and heaviness are still in his chest.

The priest takes him from the temple toward a long, low building. A shocking sight greets him along the shaded pathway: a mangy dog is licking the infected wound on the thigh of a man seated on a stone bench. Two young priests are actually encouraging the animal to go about its filthy business. The old priest explains that dogs are favored by Asklepios almost as much as snakes. The merchant looks back at the animal, which has turned its attention to its own testicles.

They arrive at the long building. This is the *abaton*, the sacred dormitory where Nikias will spend the night. It is here that Asklepios will visit him in his sleep. He may even cure Nikias then and there, though it is more likely that he will come into Nikias's dreams to tell him how he needs to be treated. This is why he must stay on for a second day, to receive the prescribed treatment. It is very important, the priest explains, that he remember every last detail of his dreams. The key to his cure may be hidden there.

The priest leads Nikias through the door. The first thing that strikes him is the smell, a thick blend of beeswax candles and musky sweat. After his eyes adjust to the darkness, he sees several dozen small beds, most of them occupied by motionless figures. And then the floor moves. More snakes—dozens of them—slithering beneath the beds. Nikias is careful not to step on them as the priest guides him to his bed. A thick fleece, made from the pelts of sacrificed animals, covers it. There is a bucket for his waste, but no drinking water. Nor will there be any food. The body must remain pure for the visitation.

It is the longest night of Nikias's life. The sacrificial fleece stinks. He is tormented by those around him—their coughs and sighs, their tossing and turning, their farts and belches. Sleep is a long time coming. Minute after minute, hour after hour, he waits for the visitation. But every time he is about to drift off, something disturbs him. Once, he is certain that a snake has slithered over his legs. There is nothing sacred about his thoughts. They are mostly of business. Or of his father's death. Perhaps he is not meant to be cured.

He manages to drift off just before dawn. He dreams that he is lying paralyzed in some weeds. His family has gathered nearby for a feast. There is laughter and music. They cannot see him. He wants to go to them but he cannot move; he wants to call out but he is mute. There is a flash of movement in the weeds—a snake's fanged mouth strikes his breast. He can feel the poison moving through him. His chest seizes like his father's did. He can feel himself falling into nothingness. He is dying.

This jolts him awake. There is an ominous pounding in his chest, and his breathing is so difficult that he feels as if that rancid fleece has been crammed down his throat. And he is more terrified than ever.

Finally the door opens and the morning sun cuts through the fetid air. The old priest quizzes Nikias about his dreams, unable to hide his disappointment as the merchant describes their content. When he is done, the priest goes to speak with his colleagues. As Nikias waits, exhausted and short-tempered, his stomach rumbling, one of the sacred dogs approaches. It is all he can do not to kick its sacred tail.

The priest returns to tell him that his dreams are not satisfactory. No cure can be gleaned from them. Another night's stay in the abaton is required. Before that, more ablutions, more sacrifices, more prayer and fasting. A further offering will also be necessary.

Nikias is no longer able to contain himself.

"I don't understand what any of this has to do with the pain in my chest."

"It is how the gods work. You must have faith."

"But isn't there something you can *do*? I'm sorry, but I'm a practical man. I see something broken, I fix it."

The old man simply stares at him.

"I do not want to die," Nikias says finally.

The priest tugs at his beard for a moment.

"There is one thing."

"What?"

The priest's eyes flicker toward the dog.

"No," Nikias says.

"Then perhaps a snake . . . "

The priest begs him to stay, but Nikias will not be dissuaded. As he steps through the gate, he remembers that his son is not due to return for another day. He must walk back to town. He wonders if he can make it. The thought of the journey wearies him infinitely. And it is not just the distance.

He knows he will be going home to die.

And then it happens, just as it had at the port. His mouth can no longer draw air; he feels a knot form in his chest. The world begins to spin and then he is falling. The next thing he knows, he is looking at the blue sky. A face appears above him— the man from yesterday, the one with the penetrating eyes.

"Can you stand?"

"I think so."

The man helps him to his feet. His hands are strong.

"So I take it your dreams were unsatisfactory," he says, his tone disdainful.

Nikias nods, wondering who this person is. The man then asks what brought him to the temple. Nikias tells him about the breathlessness, the crushing weight in his chest, his previous collapse.

"I can help you."

"But who are you?"

"A doctor."

"I've been to a doctor," Nikias says.

"There are other treatments. New treatments."

"Such as?"

The doctor glances uneasily at the temple's gateway.

"Accompany me back to town," he says. "I shouldn't stay here too much longer anyway."

"Why not?"

"Come with me and I'll explain everything."

Nikias hesitates. He is not accustomed to putting himself under the control of strangers. But he cannot make the long walk back to town. Besides, something in the doctor's manner inspires trust.

"What will this cost me?"

"Pay what you can afford." The doctor gestures to the people encamped by the weeds. "They pay nothing."

And so Nikias agrees. The doctor helps him down the many steps to his cart. As they ride back to town, Nikias tells him how he feels in more detail. The doctor responds with questions they did not ask at the temple—questions about Nikias's father and grandfather, about what he eats and how hard he works and how he sleeps.

"What is wrong with me?" Nikias finally asks.

"Here we are," the doctor answers, nodding to a large, clean building on the outskirts of town.

From the outside it looks like the abaton. Inside, however, everything is different. Large open windows admit a cool breeze. Beds line the walls, though these are unlike the reeking pestilential cots back at the temple. Some of them resemble workbenches covered with elaborate systems of ropes and pulleys. The people occupying them are calm. They are cared for by two young men. There are no snakes or dogs.

Nikias watches as the doctor attends to a man lying in one of those ingenious beds. Ropes are attached to a leather brace around his wrist; two more ropes bind his ankles to the bed's frame. The doctor carefully adjusts the brace on the arm. The man in the bed grimaces slightly, though his expression is stoic and trusting.

"His arm is broken," the doctor explains. "This device holds it in the proper position as it sets so that the patient need not end up crippled."

Nikias experiences some of the awe he was supposed to feel back at the temple. He has never seen this before: the human body is being repaired like the hull of a sailing vessel. One of the younger men brings a cup filled with a foul-smelling drink to the wounded man.

"Poppy seed and willow bark," the doctor explains. "For the pain."

As the man downs his drink, the doctor leads Nikias to an empty bed.

"Are you going to strap me down as well?" the merchant asks.

"Only if you don't follow my instructions." The doctor smiles for the first time. "Please, sit down."

The doctor places his hand on Nikias's neck and then on his wrist to check the pulse. He peers down Nikias's throat and stares into his eyes so deeply that Nikias imagines he must be able to see right into his head. Finally, his brow furrows and he nods.

"So what is it?" Nikias asks.

"Blocked vessels."

"Vessels?"

The doctor touches a spot just below the merchant's collarbone.

"Here." His hand moves to the center of Nikias's chest. "And here."

"Blocked with what?"

The doctor strokes his beard for a moment, choosing his words.

"Your body is out of balance. There is a plethora of phlegm inside your chest. It is blocking the passageways that circulate pneuma throughout your body."

"Pneuma?"

"The substance that allows you to move and think. You breathe it in and it is carried through your veins by your blood. But now phlegm blocks these vessels in your chest as they leave your heart. The heart can no longer heat the blood enough to melt away the phlegm, so the pneuma cannot get to your muscles and bones. If it gets worse, you will lose the ability to move or to think. Eventually you will die."

So it is in his heart that his trouble lies. As it was with his father. And grandfather.

"What is to be done?"

The doctor nods briskly. He is clearly a man accustomed to solving problems.

"First you must change your diet. No meat or milk. Nothing that thickens the blood. You must eat plenty of oranges and lemons. The citrus will melt away the phlegm. And you must rest."

Nikias laughs mirthlessly.

"Who has time to rest?"

"If you do not do as I say, then you will die."

This silences Nikias.

"But first we must remove as much phlegm as possible from your chest. Lie down, please."

Paraphernalia is quickly organized. A bloodstained bowl. Several small blades. One of those oddly shaped brass cups. And, ominously, a smoldering coal that the younger doctor fetches from the fire with a pair of charred tongs.

"Now," the doctor says, "this might hurt a little."

He bares Nikias's chest, then holds the brass cup upside down above the coal. After it fills with the heat, he places it on the merchant's exposed skin. It hurts, though it does not burn. The doctor explains that the heat trapped in the cup creates a vacuum that draws the phlegm out of the vessels and through the skin. As he repeats the procedure on other parts of Nikias's chest, he further explains how these fluids, which he calls humors, are responsible not just for disease, but also for emotions—anger and exhaustion and sadness. At long last, Nikias can picture what is happening inside his body. He can feel his chest opening.

After the doctor finishes, Nikias asks him about this place, about why he has never heard of it. The doctor says that he must be careful. The priests would tear it down if they could.

"Why?"

"They fear that they will lose their power once people understand that disease is caused not by vengeful gods but by things we can control."

"But would they really tear it down?"

"Priests can be stubborn." He smiles grimly. "I should know. I used to be one."

"You?"

"I wanted to be a healer."

He rubs his hands together.

"And now we must remove some of the excess blood that the blockage has caused to pool within you."

He places the bloodstained bowl on the floor beside the bed and chooses one of the small blades. He takes hold of Nikias's arm and positions it over that bowl. And then the doctor cuts him. Just like that. Without hesitation. The pain is less than Nikias feared. Blood percolates into the wound, then flows into the bowl. After a while the merchant begins to grow light-headed. A peacefulness descends. He closes his eyes as the doctor presses down on the incision. Enough blood has been drawn, he announces. Balance has been regained.

And it is true. Nikias feels better than he has in a long time. After he returns home, the new diet makes him lose weight and feel better. He works less, giving his son more responsibility, and this helps him sleep. The vision the doctor gave him of a body in balance, of a chest cleared of blockage, has allowed him to relax.

A few weeks later, he hears the news that the temple of Asklepios has been badly damaged by fire. Rumor has it that it was torched by a doctor and his acolytes, though Nikias cannot believe that the man who treated him is guilty. He is not a violent man. He is a healer. The word is that the accused has fled to Athens. Nikias recognizes his name when it is spoken around the port: Hippocrates. It is, after all, not the sort of name a man is likely to forget.

THE heart was conceived in Egypt sometime around 2000 BC. It was born fifteen hundred years later in Greece. Although other ancient cultures were awed by the cardiac muscle, the Egyptians of the Middle Kingdom were the first to attempt to understand its purpose and function. They went beyond dumbstruck wonder to try to explain this seemingly inexplicable entity in their chests. Although these inquiries were occasionally based on empirical observation, the Egyptians ultimately depended on myth and religion to tell the heart's story. For them, the heart was not of this world. It was more than the sum of its fleshy parts. It was left to the Greeks, led by Hippocrates, the father of medicine, to commence a purely scientific inquiry into the heart's anatomy and physiology, a process whose conclusions, despite their many errors, set the foundation of modern cardiology.

For the ancient Egyptians, the heart, or *ib*, was the animating core of the human body. When an Egyptian thought about the heart, it was not as just another internal organ, but rather as a pulsating slice of eternity trapped by the ribs. It was as if the heart were some sort of immortal captive just waiting to escape and return to its rightful place in the heavens. Created by a drop of the mother's blood at the moment of conception, the heart was the seat of the soul. It housed intelligence and sparked life; its incessant motion mirrored that of the ever-flowing Nile and the always shifting heavens. No other part of the body was treated with similar reverence.

For the Egyptians, the heart was the repository of a person's inner truths. It could be heavy with sin or buoyant with virtue. As such, it was the star witness on Judgment Day, its testimony determining the eternal fate of the recently deceased. *The Egyptian Book of the Dead* ordained that a dead person's heart should be left intact during mummification so it could accompany him or her into heaven. (It should be noted that usually only pharaohs, their courtiers, and the very wealthy were mummified.) During the burial process, the heart's essence—the soul—was magically transferred to a special talisman, usually a stone scarab that was balanced on the chest of the deceased. Once the soul had entered this charmed beetle, it would fly up to heaven, where it would be weighed against a divine feather belonging to the goddess Ma'at. If the heart essence proved heavier than the feather, it would be immediately devoured by the monster Ammet, in whom were combined the most fearsome elements of the crocodile, the lion, and the hippopotamus. If it weighed less than the divine feather, then the soul could proceed to the heavenly, rush-filled fields of Aaru for eternal bliss. The brain, meanwhile, enjoyed no such esteem. Deemed a fatty organ of little importance, it was removed through the nose with hooks during mummification and discarded without ceremony, as were the other internal organs. The undeclared war to determine which of these two organs ruled the human being had begun, with the first battle being won by the heart.

For some in the Middle Kingdom, the heart was not only the subject of myth. Egyptian physician-priests also attempted to study its anatomy and physiology. A group of them who worshipped the goddess Sekhmet, the lion-headed patron of healers, was known to be taking the pulse as early as 2000 BC as a way of assessing a patient's well-being. The dissection that took place as part of the mummification process provided these early researchers with ample opportunities to observe the internal body and formulate theories about how the cardiac muscle worked. The "treatise on the heart" contained in the Ebers Papyrus (ca. 1500 BC) sets out perhaps the first systematic attempt to imagine the heart's precise function within the body. In this tract, the heart is described as being at the center of a system of all-purpose vessels known as *metu*. Not surprisingly for a culture so dependent on the Nile, the human organism was imagined as a complex irrigation system, with the metu serving as aqueducts that carried urine, semen, sweat, and tears to and from various organs. The heart was this network's principal source, the place where blood was mixed with the air that was inhaled directly into the heart's chambers through the windpipe. Once infused with divine breath, the blood flooded and fertilized the body. Disease was attributed to blockages in this system; Egyptian physician-priests were enthusiastic purveyors of purgatives and emetics that were intended to keep the metu open and the body's fluids flowing unobstructed.

Despite these fledgling attempts at comprehending anatomy and physiology, there is no evidence that Egyptian thinkers were capable of imagining the heart working independently of theology. The physical functions they described were never seen as the heart's *purpose* for being. For the Egyptians it was, first and foremost, an instrument of the divine. The same was true of their rivals, the Hebrews. In both cultures the heart was understood to be the primary vehicle for godly presence in the individual.

The Hebrews, however, employed a very different system of metaphors to imagine the heart. Their thinking was strongly informed by the stern, merciless desert God they worshipped. The word *lev* (heart) appears more than eight hundred times in the Old Testament. While a few of these mentions relate crude attempts at describing its anatomy or concocting folk remedies to soothe its ills, the overwhelming majority show that the heart was seen as the touchstone of an individual's relationship with Yahweh. In fact, it is sometimes unclear whether the Hebrews meant the actual myocardial muscle when they used the term *lev*. The heart could be any part of the body, as long as it was in communion with Yahweh. The pulsations a Hebrew felt throughout his body had nothing to do with the circulation of blood. Rather, they were the reverberation of his God's word throughout his flesh.

It was in the heart that the worshipper's most sacred covenants with Yahweh were sealed. The heart was the part of the individual that followed God's commandments—or disobeyed them. The Old Testament God could be a harsh master. He demanded obedience. The heart was his means of communicating these edicts. As the psalmist says of a righteous man, "The law of his God is in his heart; none of his steps shall slide" (Psalms 37:31, King James). The wicked man, however, was the one whose heart was either unable or unwilling to follow the commandments. "He that hardeneth his heart shall fall into mischief" (Proverbs 28:14). In the Old Testament, the strong heart gives us the energy to walk with God; the weak one causes us to stumble and fall into evil ways.

Although the heart possessed a divine nature, there was nothing ethereal or abstract in the way the Hebrew prophets imagined it. It provided them with a turbulent, bloody, tactile metaphor. The Old Testament heart overflows, congeals, smolders, breaks, and quakes. "His word was in my heart as a burning fire shut up in my bones," Jeremiah says of his shame before God (20:9). This heart is also sentient. It sees what is invisible to the naked eye; it alone can hear the word of God and receive his laws. In Proverbs it is pictured as the anatomical equivalent of Moses's tablets, with Yahweh's commandments engraved upon both. If the Hebrews attempted to listen to Yahweh with their ears or comprehend him with their reason, communication would inevitably break down. Instead they had to open their hearts to him or, in the striking words of Jeremiah, "Circumcise yourselves to the Lord, and take away the foreskins of your hearts, ye men of Judah" (4:4).

This communication was not a one-way street. The heart was more than just a receiver tuned to record heavenly laws. It also generated humanity's responses to God's commandments. Prayers originated deep within the heart's chambers. The heart had a mind of its own. It was capable of a wisdom that the brain could not comprehend. It was also the crucible of the Hebrews' moral lives, the seat of a faculty that would later come to be called conscience. "Let the counsel of thine own heart stand," Ecclesiasticus asserts, "for there is no man more faithful unto thee, than it" (37:13). The wise Hebrew did not make up his mind before deciding to act; he made up his heart.

In addition to mediating a person's relationship with God, the Hebrew heart was also the seat of man's most profound, most enduring feelings toward his fellow human beings. Hearts bound us to one another. In the Jewish tradition, a man mourning the death of a parent was counseled to tear his clothes to expose the breast, a piece of graveside theater that dramatized the location of his deepest pain. Similarly, adulterous women could have their clothes forcibly torn in order to uncover cheating hearts. While the Egyptian envisioned his heart ascending to heaven on the wings of a gilded beetle, the Israelite was more likely to picture it as a stony repository of shame, obligation, and lamentation.

Along with the Egyptians and Hebrews, the Greeks also helped lay the foundation for the heart's remarkably broad and enduring use as a metaphor in Western culture. One of the earliest examples of its symbolic resonance in the Greek mind is the notion of *thumos* in the epic poems of Homer, which were probably composed in the ninth or eighth century BC. Although most contemporary scholars agree that the closest translation of thumos is "spiritedness," English translators have traditionally chosen to render it as the more resonant and poetic "heart." Even though there is no record of the Greeks of this era having a distinct conception of the myocardial organ, this translation is not, in fact, all that great a poetic liberty. Throughout both *The Iliad* and *The Odyssey*, thumos is often associated with a character's blood and breath. Whatever its physical location, thumos is the source of a person's most powerful emotions, whether they be love and devotion or anger and vengefulness. It is also where the various qualities that make up our individual characters are stored. The heart houses pride, courage, cowardice, greed, and lust. When Priam, the Trojan king, strides fearlessly among the Greeks to retrieve the body of his slain son, his enemies stand in admiration of his "iron heart." Thumos can also be understood as another early formulation of the idea of conscience, since throughout Homer's work, his heroes often enter into dialogue with their hearts, occasionally even arguing with them—or wrestling with them, as we would now say.

Perhaps the most vivid and explosive example of thumos in Homer's work can be found within the armored breast of Achilles. The great warrior's heart propels the action of *The Iliad* more powerfully than does Helen's peerless face. With Achilles, we have the first truly heart-driven character in Western literature. From the poem's first line, Achilles' heart is characterized as being full of *menin*, or wrath. This is never more evident than in his single-minded pursuit of Hector after the Trojan prince slays Achilles' friend (and perhaps lover) Patroclus. So great is the wrath boiling within Achilles that the god Poseidon refers to the hero's "murderous" heart. Achilles' heart is also capable of generating considerable lust, especially for his concubine Briseis, an emotion that causes him to abandon his fellow Greeks after she is kidnapped by their king. His heart is the seat of a pride so great that it almost costs his countrymen their victory in the long and bloody campaign against Troy.

Greek myths further rooted the heart in the Western imagination, and none more so than the one that describes the birth of Dionysus, the god of wine and dance and revelry, as told by the mythical poet Orpheus. In his account, Zeus transforms himself into a snake to make love to Persephone, the queen of the underworld, who also happens to be his daughter. The result of this incestuous union is Dionysus. Zeus intends for his son to become his heir, but this enrages a jealous Hera, who is both Zeus's wife and his older sister. She persuades the Titans, the syndicate of gods who rule the universe, to kill Dionysus. They try to do this by distracting the infant with a mirror and toys, but Dionysus, true to his nature, proves elusive, disguising himself as various animals to escape his killers. The Titans finally capture him after he takes the form of a bull. They tear him to pieces and devour him, but before they can consume his heart, Zeus incinerates them with a thunderbolt. A grieving Persephone manages to salvage this last remaining piece of her son from the flames. Later, the ashes of the Titans, along with the bull's flesh they had consumed, will be used to form humankind, which explains why our souls were thought to possess both earthly and divine qualities. Zeus, meanwhile, takes possession of Dionysus's heart from his lover-daughter and implants it in a mortal woman, Semele, who immediately becomes pregnant. Dionysus is born again, this time to rule over the dance for eternity.

And so, from the earliest expressions of the Greek imagination, the heart became a metaphor for what is most essential in a human being. Even after the rest of the flesh has been consumed, it endures. The heart is the last thing to die. In the case of Dionysus, the fact that his heart escaped being devoured allowed him to be reborn, and even to gain eternal life. From here on out, the heart will hold a place at the center of the Western imagination that no other part of a human being can rival.

The main contribution of the Greeks, however, was not in poetry or myth. It was their establishment of the Western tradition of scientific study of the heart. Beginning in the sixth century BC, pre-Socratic philosophers and physicians undertook the systematic inquiry into the heart's anatomy and physiology that continues to this day. True, healing cults such as those of the god Asklepios remained powerful through Roman times. While more and more patients like the fictional Nikias gravitated toward empirical practitioners working in the Hippocratic tradition, there were still plenty who sought relief in old-school charms and incantations. And superstition played a role in the theories of even the most scientifically inclined—the mathematician and philosopher Pythagoras was sufficiently daunted by the heart's occult power to lay down the edict *"Cor ne edito,"* or "Eat not the heart." (As far as we know, he was referring to the hearts of animals.) The prevailing impulse among the foremost Greek thinkers, however, was toward a fundamentally empirical understanding of the heart. Hippocrates made this clear in about 400 BC in his tract *On the Sacred Disease* when he called for a medical science that did not rely on the gods to explain illness. In the battle between the temple and the clinic, the clinic was gaining the upper hand. The demystification of the heart had begun.

The Greek physician understood that this organ was essential to the body's functioning—he just did not know how. He could feel its action through pulse points in the flesh; he saw that it became agitated in times of stress or exhilaration; he knew that its cessation accompanied death. Perhaps the main obstacle to progress was a deep cultural taboo that was to hamper scientists for the next two thousand years. While Greek physicians could freely examine the hearts of slaughtered beasts, human dissections were strictly forbidden. Soldiers' wounds might permit occasional glimpses into the barely fathomable netherworld, but for the most part the only way to understand the body's inner functions was through their outward manifestations. This left the doctor lost in a maze of symptoms. He could feel the heat of a fever on his patient's sweaty forehead and the racing pulse hammering in his neck, but he had no means of observing their causes. Although his impulse was to deny supernatural explanations, the body's cloak of skin kept him from cataloging the hard evidence that would allow him to send the soothsayers and shamans packing.

Because of these limitations, the Greek physician had to rely on frequently elaborate theories to explain how the heart worked. Sometimes these were inspired; often they were simply guesswork;

occasionally they were so catastrophically wrong that they sent generations of thinkers on wild-goose chases. Perhaps the most pervasive of these theories was the belief that the heart was a furnace. Because the Greek doctor could see that the body cooled when the heart stopped beating, he deduced that the heart, fueled by blood and fanned by the breath, generated the body's "innate heat," the quality that animated all creatures.

The other central theory that the Greeks developed to understand the heart was the notion of humors. Every major Greek philosopher and physician believed that the human body was made up of four humors, which were vital fluids whose balance determined the state of an individual's health. (The Greek word for humor is *chymoi*, which also means "juice" or "sap.") Blood, phlegm, choler, and black bile were combined in the human body, just as air, earth, fire, and water made up the physical world. Each of these humors was associated with particular human attributes. Fiery choler (or yellow bile) caused irascibility and mania; muddy black bile was responsible for depression; runny phlegm made us listless and was associated with colds. Blood was the healthiest of the humors, responsible for a sanguine disposition. A proper balance among the four humors meant good health. A surfeit or deficiency of any of them usually resulted in disease.

The heart stood at the center of this system. The Greek physician could not picture the cardiac muscle outside of its role in the humoral structure. He might not be able to observe its valves and ventricles in action, but he had no doubt that the organ played a role in regulating the body's vital substances. As noted in Nikias's story, Hippocrates believed that the heart was responsible for heating the blood so that it would not be coagulated by phlegm, a condition that might result in blockage, illness, and even death. "For when the veins are shut off from the air by phlegm and do not receive any," he wrote, "it makes a man speechless and mindless." He deemed the heart itself such a powerful, dominant organ, however, that he was able to categorically state, "*Cor aegrotari non potest*" ("The heart is immune to disease"). For Hippocrates and his followers (some of whom wrote books under his name), the heart was the sun that warmed the body's inner climate. When it was clouded over, illness arrived like a bleak, rainy day. When it set, the endless night of death followed.

This would explain the sort of treatment given to someone like Nikias, who suffered from coronary artery disease. Because Hippocrates correctly sensed that his patient's discomfort was being caused by a blockage in the chest's blood vessels, he did everything in his power to alleviate those symptoms. Some of his prescriptions would have helped, such as a diet high in fruits and vegetables. Bloodletting, meanwhile, might have temporarily eased high blood pressure. But nothing he did could have removed the plaque that had been gathering in the lumen—the interior cavity—of the merchant's coronary arteries. Nikias may have had a few good months after the doctor's intervention, but his long-term prognosis remained poor. He was simply born too early to survive his disease. More than two thousand years would have to pass before the doctor's intellectual descendants would be able to prevent blockages with cholesterol-lowering drugs, or relieve them with bypass surgery or balloon angioplasty.

Although Greek thinkers were largely unified when it came to the heart's importance in regulating humors, they were divided into two distinct camps over another key question—whether the heart could think. As difficult as this may be for the modern reader to conceptualize, many Greeks saw the heart as the seat of consciousness and intelligence, where our dreams and memories and volition were housed. Other philosophers considered the brain to be the thinking organ. This debate raged until the third century BC, when Greek physicians working in Alexandria, where human dissection was permitted, discovered the nervous system. But in the realm of metaphor, the heart-versus-head debate has continued right up to our time. The brain might stand at the center of our nervous system, but the heart remains a powerful symbol of those qualities that go beyond rational thought: intuition, emotion, and faith.

Pythagoras (ca. 580–ca. 500 BC) was the first Greek thinker to leave a record of his views on the heart. While he placed a human being's rational capacity in the brain, he carried on the tradition started by Homer when he suggested that the heart was the wellspring of emotion. In the fifth century BC, his follower Alcmaeon of Croton expanded on his master's theories when he originated the concept of pneuma, which would be central to all discussions of the heart for the next two thousand years. Pneuma, defined roughly as divine breath, was an invisible atmospheric substance, a sort of rarefied air that enabled thought and sensation. Pneuma was intelligence. It was what made the heavens move in an orderly fashion; it was what caused two plus two to equal four. By breathing it in, human beings gained the capacity to think and move and feel. They swam in an invisible sea of thought. In Alcmaeon's view, pneuma was inhaled directly into the brain, from where it was sent to the limbs and organs to perform the mind's will.

He was opposed in this by the philosopher Empedocles (ca. 490–ca. 430 BC), who was also a follower of Pythagoras but who broke with his master to give the heart pride of place in the human organism. He saw it as a sort of thinking engine. In Empedocles' view, breathing brought pneuma directly into the heart, enabling it to think and to will activities such as movement and speech. Breathing cooled this hot engine, keeping it from overheating. This explained why the athlete gasped for breath after a race and the soldier's chest heaved during battle. They were not only replenishing the pneuma required to perform these stressful activities, but also keeping their engines from overheating. Sleep, meanwhile, occurred when the furnace slowed and the body's temperature lowered. Death was simply the natural conclusion of this cooling process.

Plato (ca. 428–ca. 348 BC) may have called Empedocles his "gentle muse," but he broke from his predecessor by returning the seat of intelligence to the brain. Like all of the first great philosopher's theories about the physical world, Plato derived his views on the heart from metaphysical thinking rather than empirical study. Plato believed that the soul was divided into three parts: the immortal, the mortal, and the appetitive. The immortal soul, which he believed was capable of a type of reason that could understand eternal truths, was located in the head, since this was closest to the gods and farthest from the churning, malodorous business of the stomach and bowels. The mortal soul, that turbulent locus of "terrible and irresistible affections" like fear, anger, greed, and pleasure, was situated in the chest, while the base appetites such as hunger and thirst were housed in the belly. The heart, therefore, was deeply implicated in the doings of our mortal souls and in those irresistible affections. It was a fountain of feeling: Plato pictured the heart as a "knot of veins" that gushed blood in moments of passion or "when danger is foreseen." When this happened, Plato did indeed follow Empedocles by claiming in his *Timaeus* that the lungs kicked into action.

> But the gods, foreknowing that the palpitation of the heart in expectation of danger and the swelling and excitement of passion was caused by fire, formed and implanted as a supporter to the heart the lung, which was in the first place, soft and bloodless, and also had within hollows like the pores of a sponge, in order that by receiving the breath and the drink, it might give coolness and the power of respiration and alleviate the heat.

Although Plato was too interested in making the body conform to his metaphysics to be of any use as an anatomist, his vision of the heart as the wellspring of passion was an important step in its deepening metaphorical identity. While Greek poets such as Homer and philosophers like Pythagoras had suggested that our most intense feelings emanated from the heart, it was Plato who most coherently (and influentially) stated that this hot, pulsating, excitable organ was the engine of our anger and pride, our courage and sorrow. It could also be the source of erotic desire, though not of love in its highest sense. For Plato, the quickening of the heart that occurred when a person saw his or her loved one was just a step in the ascent to true love, which could happen only in the mind, after the lover comprehended what was eternally true and beautiful in the beloved. Platonic love existed beyond all the blood and heat contained in the heart. This split between passion and piety, between lust and love, would resonate throughout the Middle Ages and the Renaissance, and it continues up to the present day.

Plato's prize student, Aristotle (384–322 BC), in keeping with the Greek tradition of spirited debate, broke with his teacher's vision of a brain-dominated body and relocated the seat of intelligence and sensation in the heart. He did this by relying on a more empirical approach than Plato used (which is ironic, given that Aristotle was so very wrong). The son of a doctor, he was an enthusiastic anatomist of animals who observed that the heart was the center of the body's network of vessels and must therefore be its most vital organ. After watching a heart starting to beat in an embryonic chick, he claimed that it was the prime mover of all life, since it was the "first to live and the last to die."

In Aristotle's conception, life was all about heat. The process began when food was cooked in the stomach and bowels to form a bloodlike substance. Once brewed, it boiled up into the heart, where it mixed with pneuma. From there, further heating moved it throughout the body to deliver pneuma to the muscles, bones, organs, and skin, thereby enabling motion and sensation. Thus, for Aristotle, the body's blood vessels served as nerves. The brain, meanwhile, was little more than a phlegm-producing cooling system for the heart.

Although Aristotle understood that the heart distributed blood throughout the body, he had no concept of circulation (nor would anyone else for two thousand years). Blood was sent out from the heart, but it did not return. It was completely consumed by the hungry organs and tissues. Nor did he comprehend that a heartbeat produced propulsion. For him, pulsation was caused by blood boiling in the heart. Aristotle was also hindered by some key physiological mistakes, such as his postulations that the heart contained a third ventricle and a bone.

Taboos against human dissection presented no problem for two great physicians who worked just after Aristotle. Herophilus and Erasistratus were Greeks, but they practiced in the Egyptian city of Alexandria in the third century BC, where they were able to perform human autopsies and may even have vivisected condemned criminals. For the first time, scientists working in the Greek tradition were able to examine the actual human heart—perhaps even while it was still beating. Not surprisingly, they gained a deeper understanding of the organ than their predecessors had. After observing that a corpse's veins would collapse when drained of blood but the arteries retained their suppleness, Herophilus differentiated between the two types of vessels, dividing the vascular system for the first time. In his view, veins carried only blood to the body's muscles and organs, while arteries transported pneuma as well as blood. He was also able to observe the separate nervous system, tracing it from its origin at the brain stem and throughout the body. Aristotle's belief that the heart was the origin of intelligent action was thrown into doubt. Given the view gained from the body's increasingly transparent structure, it came to be considered much more likely that the brain was the source of willed movement.

Erasistratus, who was Herophilus's pupil, provided an even more comprehensive picture of the heart. Taking his cue from his master's separation of the functions of veins and arteries, he envisioned the heart as a sort of double wellspring whose right side sent nourishing blood to the organs and extremities via the veins while the left side performed the same function with pneuma. Although he had no way of knowing it, this view of the heart performing two distinct functions came tantalizingly close to explaining the circulation of the blood. However, Erasistratus's Aristotelian belief that blood was completely consumed by muscles and organs prevented him from making the breakthrough that would have to wait nearly two thousand years more to occur.

And then came Galen. Hellenic cardiology reached its clearest, most detailed expression with this Greek-born, Rome-based physician who flourished in the second century AD. A bold clinician and prolific writer, Galen set forth a vision of the human heart that was comprehensive, lucid, and, in several key areas, completely wrong. But such was the power of Galen's theories that it was not until the sixteenth century that physicians were able to escape his formidable spell.

Galen's view of the movement of the heart and the blood can be likened to a process of distillation. As he imagined it, the heart's action began in the liver. It was there that food transported from the stomach and intestines was brewed into a thick, sludgy substance that contained what he called natural, or nutritive, spirits. Most of this solution percolated out to the organs and the extremities, where it fed muscle and bone. Some of it, however, was piped to the right chamber of the heart, where it was further cooked by innate heat into a thinner, lighter substance. This rarefied blood was transported into the left ventricle through invisible pores in the septum, the wall that divides the heart into right and left halves. In the heart's left chamber, blood was combined with inspired pneuma. Infused now with the vital spirits that made basic movement possible, this bright red blood coursed through the arteries to the body's muscles. Arteries also returned waste products from the heart to the lungs, from which they were exhaled. Finally, a small amount of blood traveled to the brain via the carotid artery, where it was further refined with something that Galen called "psychic" spirits. This further-distilled blood was then sent to the body in order to enable willed movement. Vital spirits, Galen believed, gave our fingers the ability to twitch; psychic spirits enabled them to grip a pen and write legibly.

Galen did not understand that blood circulated. As with all Greek physicians, for him its movement through the body was along a one-way street. Another key shortcoming was his failure to understand that the heart is a pump. For Galen, the purpose of the heart's beating was not to propel blood. Rather, he saw the heart as a bellows, dilating in order to suck in cooling, vivifying air from the lungs. Contraction, meanwhile, expelled a waste product created when pneuma and blood mixed. What moved blood in the Galenic body was a vaguely described combination of muscular contraction in the thorax, a pulsing that occurred within the arteries themselves, and the attractive pull of hungry muscles and organs. Behind it all was the general tidal movement of the body's humoral system, where the four chymoi were in a constant state of flux, attempting to achieve the balance that would enable good health.

After Galen, Greco-Roman inquiry into the heart effectively ended. His theories were considered the last word on the topic. For the next thousand years, doctors and scientists, almost without exception, believed that his view of the heart comprehensively explained its form and function. And yet, despite having been given the passionate attention of some of history's greatest minds, the heart remained almost as much of a mystery to the Greeks as it had been to the Egyptians and Hebrews. Although Greek philosophers and physicians had largely removed the gods from their explanations relating to this sublime engine, their understanding of physiology remained in several key respects just as fanciful as that of the Egyptian priests and the Hebrew prophets who preceded them. From Plato to Galen, the inability to observe the human heart in motion meant that speculation was needed to posit a comprehensive picture of its function. Fictional substances like pneuma and humors were created to explain symptoms whose true causes could not be observed. Veins and arteries were indeed differentiated, and mental processes were rightly situated in the brain. But the organ's function as a pump and the consequent circulation of blood—the heart of the heart's action—were not yet understood. Nor would they be for a very long time. The Dark Ages loomed, during which Greek efforts to demystify the heart were put on hold and its mythic role once again reigned. It would be well over a thousand years before Renaissance physicians were again able to explore the heart's anatomy and physiology with anything like the scientific rigor employed by their Greek predecessors.

CHAPTER 2

Sacred Heart

Umbria, Italy, 1308

He does not want to travel into the Umbrian hills. It is August, the hottest time of the year. Berengario can almost feel the punishing sun and taste the parched dust as he contemplates making the difficult journey to Montefalco. But he has no choice. The bishop has ordered him to leave immediately. He has never seen his superior so indignant. From the moment word had arrived of a miraculous heart at the convent, Trinci had been deeply suspicious that the nuns were perpetrating some sort of hoax. It would not be the first time that a group of holy sisters had got themselves up to such mischief, he'd said darkly. If that were true, then their heresy would need to be suppressed as quickly and as absolutely as possible.

And so Berengario is once again cast into the role of inquisitor. He leaves at dawn, perched on the donkey that has taken him on so many of these journeys. He travels alone, though of course he is not alone. The full weight of the church is with him, and it is an awesome authority that could crush the entire convent if it is discovered that this miracle is in fact a sham. Although Berengario's official title is vicar to the bishop of Spoleto, conducting inquisitions is becoming his sole occupation. And that means he is a busy man. The world seems to be filled with fakes and heretics these days, all of them doing Satan's work.

He has certainly uncovered more than his share of hoaxes in the past ten years. They seem to be especially rampant at convents, and they often involve the heart. More than once he has discovered that stigmata appearing on the breast of a nun, claimed to be the external manifestations of a holy heart, were self-inflicted. Recently he had been summoned to a convent after a newly deceased sister was discovered to have no heart at all, supposedly confirming her claim that it had been plucked from her chest by Jesus Christ and taken to heaven some five years earlier. After closely interrogating a trembling novice, Berengario found the shriveled organ freshly buried in an orchard just beyond the convent's walls. Severe punishment was meted out for that heresy: the convent was closed, and several of the sisters were burned at the stake. Berengario took no pride or pleasure in this.

Of course, he is also capable of mercy when the situation calls for it. Often, a claimant was not guilty of any deliberate deceit. Raptures and visions could often be ascribed to the delirium that resulted from fasting or the loss of blood occurring with self-mortification. This is something the bishop and others in the church do not seem to understand. There is not always malice in the hearts of those who make false claims about miracles and visions. Still, the lies have to be exposed. Just because someone lacks malice does not prevent that person from doing Satan's work. A false claim of a miracle was a serious sin that needed correction.

A claim could also be true. He must bear this in mind as well. Twelve centuries earlier, Saint Ignatius, one of the church's first martyrs, was found to have an impression of the crucifixion within his heart after having the organ ripped from his chest by a pagan king, who immediately accepted the Lord as his savior. Recently, in Città di Castello, a nun's heart was found to contain perfect images of the three members of the holy family, each one in a little glass ball. There could be no doubt about these things. They had been confirmed by the church. He must keep an open mind.

As Berengario's donkey kicks up dust on the winding mountain road, he goes over what little he knows about recent events in Montefalco. Sister Chiara, the nunnery's abbess, had died two weeks earlier, and while her body was being prepared for burial, perfect copies of the instruments of Jesus's Passion on the cross were discovered embedded in her heart. Although talk of a miracle quickly swept through the region, some members of the local clergy had complained to Bishop Trinci, implying that Chiara's fellow nuns were perpetrating a ruse. For now, the heart in question was being kept under lock and key. It is up to Berengario to determine whether it is indeed a holy relic and Chiara is a candidate for beatification.

Or whether someone should burn.

He arrives in the small town at midday. Straight away, there is evidence that extraordinary events are happening. The main square in front of the church is packed with people, even though the sun beats mercilessly down upon them. As Berengario passes through the crowd, he can tell that many are sick or crippled. He has seen the expressions on their faces time and time again. It is how people look when they are expecting a miracle. This is why he is here. There is nothing more painful to him than the dashed hopes of a suffering believer.

Although he is hot and tired, he goes straight to the convent of Santa Croce, which is situated in a walled compound shaded by a grove of tall cypress trees. The crowd is even thicker here—Berengario covers his mouth to endure the stench. The portress is clearly terrified when she peers through the grate and sees that the inquisitor has arrived. He demands to be escorted directly to the new abbess. Her name is Sister Johanna. She offers to take him to the oratory to examine the heart, but that is not how he conducts his inquisitions. First he will need to learn about all of the circumstances surrounding the alleged miracle. He has to understand what he is supposed to be looking at before he actually casts his gaze upon it. He will also need to pray. The devil is very good at tricking the eyes and seducing the imagination. The understanding that the Lord puts in your heart, however, is the one thing that remains beyond evil's reach.

He can see immediately that the abbess is no fabricator. She explains recent events clearly and without prejudice. Chiara, the former abbess, died fifteen days earlier. She was forty years old. Popular and beloved, she had suffered pains in her chest for almost fifteen years, ever since she had had an extraordinary vision. Johanna had been a novice at the convent when the visitation had happened; her memory of it was still clear after all those years. It had come during the celebration of the Epiphany, just after Chiara had become abbess. She had collapsed during the ceremony, the golden ampoule she was holding rattling across the stone floor. Although at first everyone had thought she had simply fainted, her trance had lasted three weeks. It had been all the nuns could do to keep her alive, forcing sugar water between her parched lips and taking part in a prayer chain that lasted day and night straight through those endless three weeks.

When Chiara finally had awakened, she reported that she had been visited by Jesus at the moment of her collapse. He had looked sad and exhausted, dressed in tattered clothes and struggling beneath the weight of the massive splintered wooden cross he carried. He had told her that he had been wandering since his crucifixion and had not yet found a place to plant his burden. And so Chiara had offered to take it from him. He responded by planting the cross directly inside her heart. The pain was like nothing she had ever felt. The weight had crushed her ribs; sharp splinters had pierced deep into her chest's flesh. Only the knowledge that she was bearing the Lord's burden had made the agony endurable.

According to Johanna, Chiara had told the story often since then, which makes Berengario wonder if she might have suffered from the sin of pride. Chiara had also complained regularly about the splintering pain of the cross, which suggests to the inquisitor that she might also have been a hysteric. And yet, because she had been so devout and revered within the convent, no one had ever doubted her. Local people had come to her to be blessed. There were even stories of miraculous cures.

Her body was placed in the oratory for five days after her death, the abbess explains. Even though it was the hottest time of the year in the Umbrian hills, the flesh did not decay, and the usual foul smell of corruption did not issue from it. Despite these extraordinary circumstances, Johanna finally ordered that Chiara be buried. The abbess did not want to create any more disturbance. Better to put the beloved sister to rest and get on with the regular business of the nunnery.

The body was prepared for burial by Francesca, an aged nun who saw to the daily medical needs of the nuns. As was customary, she first eviscerated the corpse, removing all the internal organs through a wound they opened in Chiara's back. These were buried immediately, as was also customary, though because of Chiara's holiness and her peculiar history, Francesca requested permission to place the heart in its own casket for preservation. Johanna agreed. The body was then embalmed with balsam and myrrh.

It was later that evening that the alleged miracle was discovered. Francesca and some of the older nuns had returned to the oratory after vespers. They removed the heart from its box and cut it open. What they found confirmed what Chiara had been saying all along: the instruments of the Passion were visible in her heart. The most conspicuous item was the crucifix. It was perfectly formed, with the body of the Savior plainly visible upon it. There was also a scourge whose fronds were tipped with metal balls and jagged bones, a crown of thorns, and three rusty nails.

Berengario asks whether Francesca had permission for such an extreme action. The abbess hesitates before explaining that the older sister claims that she had considered inspection of the heart to be included in her permission to remove it and store it separately for veneration. Johanna admits that she had not expressly forbidden Francesca from inspecting the heart more closely. Although Berengario does not like the sound of this, he decides not to press the point. If the old nun is to be punished, it will not be for a small act of disobedience.

After this astonishing discovery had been made, Johanna was summoned immediately to the oratory. It was true—the instruments of the Passion were clearly visible in the heart. Word spread quickly through the convent and beyond. Chiara's brother, a friar at the local monastery, and Simon, the doctor who had been present at her death, were summoned. They, too, confirmed that features evoking these items were woven into the heart's bloody fabric. A miracle had indeed occurred.

Chiara's heart was taken to the adjacent church in order to be properly venerated. Crowds gathered. Before long, the sick and the lame were brought forward so their bodies might be healed. But not everyone was happy about the course of events. Some members of the local clergy, including the convent's own chaplain, expressed serious doubts about the miracle and claimed that displaying the heart without proper authorization risked charges of heresy. Worried now, Johanna quickly agreed to place it under lock and key until the bishop could be consulted.

Berengario thanks the abbess, convinced that she has played no role in the fraud—*if* there is a fraud. Next, he summons the friar and the doctor who was present at Chiara's death. Both affirm that they have inspected the heart and consider it to be as miraculous as has been described. Berengario asks whether they were in attendance when Francesca removed the heart or when it was later cut open and the instruments of the Passion were discovered. Both admit to being absent on these occasions. The doctor, understanding what the inquisitor is getting at, claims that in his experience it is highly unlikely that anyone could manipulate the heart to make these things appear. To do so would require some sort of witchcraft.

The next interview is with John Pulicinis, the convent's chaplain and the man whose complaint to the bishop set the inquisition in motion. He confirms that there is indeed something in the heart that looks like a crucifix. The problem he sees is in its unknown origin. Everyone knows that Satan is capable of placing false stigmata on the body. He begins to cite examples, but Berengario cuts him off. He is well aware of these cases—he presided at several of them. Pulicinis argues that Chiara was by no means as devout as her sisters claim. She was, as Berengario suspected, a prideful woman who had constantly spoken of her visions and her suffering, wearing them like a woman in the outside world might wear jewelry or a precious garment. And she had been something of a glutton, given to eating large quantities of the best food, which she would conspicuously purge afterward. She had also been quick to take expensive medications that she had Francesca secure from the apothecary. The implication is clear. If the instruments of the Passion are indeed in her heart, they may have been placed there not by the Lord but rather by the same dark force that had tempted her to gluttony and pride.

The chaplain's doubts are echoed by a friar from the nearby monastery, Tomasso Boni, who adds several accusations of his own. According to him, Francesca is more than just a wise old nun who can mix potions to help the sisters sleep. She is a remarkably skilled healer who is fully capable of manipulating the heart—especially with the aid of Satan. The fact that she had carried out her initial examination under the cover of darkness should be taken into account. The inquisitor assures him that it most certainly will.

He asks to be taken directly to Francesca's cell in the hope that he might surprise her. But she clearly expects him—her sisters must have warned her that he was coming. Either that, or she possesses unholy powers of perception. She is an old woman, her face lined and her eyes watery, though he can sense her sly intelligence. He has known nuns like this, outwardly devout but filled with the wiliness the Lord gave Eve. He looks around the cell, which is crammed with roots and herbs and jars. There are scrolls and books as well. He remembers what the doctor said about the skill that would be required to manipulate a heart.

He questions her closely, but her answers are all correct and proper, as he suspected they would be. Although she shows Berengario the respect he deserves, he cannot help wondering if she is taunting him. She claims to have prepared the body as she would that of any nun; she separated the heart only after receiving permission from the abbess. As for its later inspection, it was carried out at night to avoid the heat and the flies. She insists that she had authorization, or at least believed she did. After all, why else would she accord the heart special treatment in the first place, unless it was miraculous? Cutting it open was merely another step in the process.

Berengario concludes their interview with a sense of frustration. Usually, by this stage of an inquisition, he knows whether fraud was involved. But now he has no idea what is the truth. The only thing left to do is to view Chiara's heart.

Before being taken to the oratory, he walks to the nearby church to pray. His own heart must be as clear and pure and filled with God as possible before he sees the nun's. He kneels before the altar in the nearly empty church. His prayers help settle his mind, but bring him no illumination. When he is done, he sits in a pew and looks up at the image of the crucified Christ. He pays particular attention to the blood running from his wounds, especially the one just beneath his ribs. The problem is that he cannot picture a cross lodged in a woman's heart. For fourteen years she was supposed to have borne nails and thorns in her flesh. He thinks about his own heart. Although he had studied Aristotle's teachings when he was at university, Berengario has his own ideas about how the heart looks when it is within the breast of a believer—ideas that were confirmed the few times he had seen the organ after evisceration in preparation for entombment. For him it is a sort of third ear, an organ that can hear the word of God. It listens in a way that a man's actual ears never could. The true Christian feels his heart quicken when God speaks to him; he senses its spreading warmth when understanding comes to him. Berengario places a hand on his ribs and feels his heart beating. On and on. Who commands it to do this? It must be the Lord. And, Berengario realizes, if he can cause such a miracle to happen, then surely he could place these instruments inside a woman's ribs. Just because Berengario's mind cannot grasp this does not mean it is false.

Feeling his beating heart, he knows what he must do. The testimony of human beings will remain baffling and contradictory. There are many more voices to be heard before he can file his report. Theologians, doctors, and lawyers. Priests and monks. Some of them will have to travel for several days to get here. Each will be influenced by his own motives. Some will be looking for a miracle, and others for a woman to burn. He will listen and he will carefully note what every last one of them has to say. But he will know all the while that the only way of being certain of what is true is to put aside the human voices and listen to the still, eternal voice in his heart.

So resolved, he strides back to the nunnery through the

gathering dusk and asks to be admitted immediately to the oratory. Johanna and Francesca offer to accompany him, but he tells
them that he must do this alone. The room is darker and cooler
than the rest of the nunnery. It feels different in here. There is
something otherworldly about this place. He picks up a candle
and moves to Chiara's shrunken, shrouded body. It rests on a
stone plinth. As always, he is struck by how fragile the dead look,
how weak and defenseless. It is a reminder of what sorry sinners
we all are. Beside the body is a small casket. He crosses himself
and prays over the body for a moment before turning his attention to the container. After hesitating for one more moment, he
lifts off the lid and holds up the candle to reveal the contents.
And there it is. Her heart. A human heart. It, too, looks paltry. A
gray, shriveled, shapeless mass. Something a butcher might toss
aside as offal.

He holds the candle closer so that its pale rays can illuminate
the shadowy folds of the recently cut organ. And he sees it. A
crucifix. There is no mistaking it lying across the heart at a slight
angle. It is almost two inches long. True to Chiara's dream, the
base of the cross is embedded right in the flesh, as if it was driven
in. The Savior's body is clearly visible upon it, his skin white
except for the bright red wound in his side, where the Roman
soldier lanced his heart. The scourge that tore his flesh rests at
the crucifix's planted base; the crown of thorns floats above his
head. Berengario even sees the nails, three of them, bent and
twisted from having been driven through his flesh and bone.

The room suddenly becomes still and silent. The passage of
time arrests; space closes in around him until it is a solid thing.
And he knows. Something that the doctors and monks and priests

could never tell him. He can feel it, here and now, in his own heart. The Lord's presence. The same presence he felt while looking at the crucifix back at the church. Suddenly Berengario can see Jesus standing before this poor woman; he can feel the pain she endured for all those years. There is no doubting it now. Sister Chiara had indeed been carrying the instruments of the Passion within her.

Light-headed with fear and wonder, the inquisitor drops to his knees and begins to pray. His own heart thunders. He feels God's presence in the very fibers of his own body. He can feel Jesus's warm spilled blood flowing through his own body. When he finally rises to his feet, he looks once more at the nun's heart. And he knows that it is true.

DURING the Middle Ages—the years of approximately 500 to 1500—the heart might very well have vanished. It was, after all, a time when little effort was made in the West to probe the mysteries of the human body. The mortal frame was an object of fear and contempt rather than curiosity and awe. The rise of Christianity in the fourth century initiated a thousand years of stagnation in the disciplines of anatomy and physiology. The body came to be seen as a pestilential, sinning vessel in which the soul was trapped for the duration of our short, brutish journey toward eternity. Medicine became a loose patchwork in which folk remedies, superstition, and quackery often took precedence over empirical methods. Doctors treated symptoms, sometimes quite effectively, but because disease was seen to be a punishment for sinful behavior (or some other act of divine will), there was no reason to look further for causes. The medical theories of the masters of antiquity such as Aristotle and Galen provided what systematic framework was required. As far as medicine was concerned, it truly was a dark age.

This was especially true when it came to the heart. Although skin ailments and broken bones might at least be treated and occasionally alleviated with rudimentary care, there was no possibility of what we now know to be heart disease receiving any sort of effective treatment. Cupping, dietary changes, and venesection (bloodletting), the therapies Hippocrates employed in Nikias's case, might occasionally have temporarily relieved symptoms, but these palliatives did nothing to address actual disorders. Perhaps the only reason that little evidence can be found of heart disease during this time is that the healthier diets and less sedentary lifestyles prevented it from developing before people died of other causes. Ironically, the Black Death probably kept more people from having heart attacks than all the statin drugs and exercise regimens in the world.

It was an era when the body was no longer an object of beauty and veneration. The idealized figures of Greek and Roman statuary gave way to the drab, two-dimensional religious pilgrims with bad haircuts and ill-fitting clothes who populated frescoes and woodcuttings. Leonardo da Vinci's proportionally perfect Vitruvian Man and Sandro Botticelli's radiantly naked Venus, created in 1490 and 1485, respectively, were still a long way from being born. Christ's scourged, resurrected corpus became the true object of bodily fascination. His was the sole beautiful flesh.

The Roman Catholic Church bore primary responsibility for the lack of medical advancement. Its efforts to stamp out pagan religions often set it at odds with traditional healers. Although many of these healers included elements of the practice of magic in their remedies, the more empirically minded among them undoubtedly were also persecuted. The priest outranked the doctor at the patient's bedside; prayer was the chief prescription for all ailments. Anyone who challenged the church's authority in matters of medical theory ran the risk of being charged with heresy, just as surely as an astronomer did for envisioning the heavens anew.

True, some within the Catholic Church worked hard to care for the sick. Monasteries often included hospices—precursors to modern hospitals—where people suffering from acute illnesses such as leprosy and even plague were looked after. Many of these same monasteries also contained gardens where medicinal herbs were cultivated. But this work involved treating the sick, not looking into the causes of disease or mapping human anatomy. The epoch's best thinkers attended to the individual's spiritual well-being instead. Theology trumped biology. For most of the Middle Ages, the only institution resembling a medical school was in Salerno, in southwest Italy. Perhaps the most popular book on what we now call health care was *Ars Moriendi*, or *The Art of Dying*, a grim self-help manual from the late Middle Ages that gave the reader practical advice on how to leave the world in a dignified, devout manner.

Another illustrative text was the *Leechbook of Bald*, compiled in England in the ninth or tenth century. (*Leech* does not refer to the worm used to bleed a person but, rather, is Old English for "healer.") A compendium of contemporary treatments for a wide range of illnesses, it provides a graphic picture of medieval thinking on the care of the human body. There is almost nothing in it about the causes of disease. Remedies primarily involve herbal mixtures, charms, and incantations. Although the substances used, such as nettles and honey, are usually harmless and may even have some crude medicinal value, they are occasionally alarmingly invasive. For instance, the "warm thin ordure of man"—excrement—is proposed as a balm for a rash, while the recommended treatment for ringworm is ground-up glass that is placed in a lesion for the parasite to fatally ingest (as opposed to being absorbed into the patient's bloodstream, it is hoped). There is very little said about the heart, and nothing at all about taking a pulse.

In the Western world during this time, the theories of Aristotle and Galen were considered the last word in human anatomy, so Arab physicians were often the ones to advance the cause of medicine. The brilliant thirteenth-century Syrian Ibn al-Nafis was perhaps the most notable among them. He was able to disprove Galen's belief that there were holes in the heart's septum, a breakthrough that prefigured Sir William Harvey's discovery of the circulation of the blood four hundred years later. An earlier great Arab doctor was Abulcasis (936–1013), who lived in Moorish Spain. Considered the era's greatest surgeon, he is credited with inventing close to two hundred surgical devices, including forceps for delivering babies and catgut for stitching internal wounds.

And yet, even in these dark ages of medicine, the heart was not forgotten in the larger culture. It found a way to remain at the center of discourse. While it is safe to say that precious little thought was given to the kidneys or the pancreas or even the brain during Christianity's first millennium and a half, the heart's hold on the human imagination deepened. The reason for this was simple—it was seen as the part of the body in which Jesus Christ dwelled. People might not have cared too much about how the heart muscle functioned, but they certainly thought a great deal about its spiritual capabilities.

The centrality of the heart in Christian worship found one of its earliest and most powerful advocates in Saint Augustine of Hippo. Born in North Africa in the fourth century and raised as a Christian, Augustine went astray in his early adult years, following a dualistic religion and living with a concubine, with whom he had a son. He was, by his own account, a man of strong passions who had a powerful interest in worldly delights. Then, in his early thirties, while working as a teacher of rhetoric in Italy, he underwent a conversion to Christianity. For Augustine this was, above all else, an affair of the heart. After hearing a childlike voice urging him to "take up and read," he opened the Gospels at random and discovered a single sentence that changed him forever. "For instantly at the end of this sentence," Augustine wrote, "by a light as it were of serenity infused into my heart, all the darkness of doubt vanished away."

Augustine made the conversion experience for Western man a matter of the heart. In his subsequent writings, particularly his autobiographical *Confessions*, he established the heart as the one true way to Christ. He introduced the idea of the *cor inquietum*—the restless heart—that felt at home only when it took up residence in the Lord. "You have made us for yourself, O Lord, and our hearts are restless until they rest in you," he claims in what is perhaps the book's most famous passage. With Augustine, the Christianized heart became a sort of safe haven within a corrupt republic of flesh, a sanctuary to which the blindly wandering soul returned after stumbling through the snares and pits of a treacherous world. After his death, Augustine would be associated with the sacred heart, the popular Catholic icon that depicts Christ's flaming heart pierced by an arrow. During his life, however, Augustine's metaphorical use of the heart was both simple and persuasive. If the lonely sinner wanted to be saved by Christ, the place to do it was in the four-chambered dwelling that rested inside his ribs.

Over the next ten centuries, it was a vision that was to become more than just a metaphor. To Sister Chiara (Saint Clare, who was canonized in 1881, five centuries after her death) and believers like her, this image of the heart as the Lord's earthly workshop, the place where he forged the Christian soul, was deemed to be literally true. Belief in his cardiac presence often found extreme expression, such as in the case of Henry Suso, a fourteenth-century mystic who used a stylus to carve Christ's name so deeply into his chest that the scar remained visible there for the remainder of his life. A German nun also reportedly engraved the shape of the cross into her left breast and then repeatedly scraped open the wound to prevent it from healing. This self-mutilation might be shocking to us today, but it should be remembered that it was during this time that a pope first decreed that the wine a communicant drank was not merely transubstantiated blood—it literally *was* the blood that flowed from the crucified Christ's heart after it had been pierced by the lance of the Roman soldier.

Of course, the instruments in Chiara's heart need not have been put there by God—or by the devil. Instead, with the benefit of hindsight, we can see that her symptoms and the subsequent physical findings—like those of so many miracles over the ages— have a plausible explanation that the inquisitor never would have had the chance to consider. In Chiara's case, there is a strong possibility that she suffered from rheumatic fever as a young woman. This illness was once common in young people and usu- ally followed a simple strep throat infection. Antibiotics have all but eliminated it in Western society today, but the young nun would not have had access to such drugs. The acute illness pro- duces fever, chest pain from myocarditis (inflammation of the heart muscle), joint inflammation, a rash, and involuntary move- ments of the limbs. One can easily see how this might be con- fused with rapture. The myocarditis and chest pain can recur, eventually leading to thickening and calcification of the mitral and tricuspid valves. When the front of the heart is removed, the crucifix-like shape of the mitral annulus and interventricular septum becomes apparent. As was so often the case, the heart's anatomical complexity and vitality would have provided behold- ers with all the evidence necessary to see exactly what they wanted to see.

If the heart was seen as a godly residence, Jesus Christ did not live there alone. An individual's soul was present as well— literally. For the Christian in the Middle Ages, whether he was an uneducated serf or a scholarly bishop, the idea that our bod- ies housed our eternal souls was also more than just a metaphor. The soul was a specific physical object. As such, it needed a place to rest during its sojourn on earth. That place was the heart. The irrevocable bodily death that occurred when the heart stopped beating was caused by the soul's relocation to a new abode, one with much more space and a far better view—or, in the case of a sinner, a squalid dive where the heating was always on way too high.

The heart was also seen as a book. It was the soul's manuscript. As the scholar Eric Jager points out in his excellent history of the subject, *The Book of the Heart*, the medieval imagination often held that a believer's every word and action were in reality inscribed directly into the fleshy tablets of his or her heart. In paintings from this era, the resurrected dead are pictured with open books detailing the days of their lives sprouting directly from their chests. A sinner's ability to deceive his fellow man, to hide "evil thoughts, murders, adulteries, fornications, thefts, false witness [and] blasphemies," as Matthew laments (15:18–19), stemmed from the fact that this written record of his true intentions was hidden from the rest of the world by a wall of flesh and bone.

This book of the heart had another author as well: God. It was his means of communication with the believer. It was the soul's instruction manual. This is what Paul meant in his second letter to the Corinthians when he claimed that "the epistle of Christ ministered by us, written not with ink, but with the Spirit of the living God; not in tables of stone, but in fleshy tables of the heart" (3:3). A sinner had no excuses. The Lord's words were tattooed on a part of his body that could be neither removed nor ignored. Even though he could not see it with his eyes, he was capable of reading the heart's inscriptions through worship and prayer, a spiritual literacy imbued by God.

In addition to being a book, the heart was also seen as a sort of alternative mind. Medieval thinkers followed Galen's theory that our everyday mental functions, the reflexes and movements of our nervous systems, stemmed from the brain. But they also posited another sort of thinking, this one situated in the heart. The brain that we used to negotiate our daily lives would never be able to know God truly. For this we needed a special acuity, a different way of understanding. This is why Augustine and others located the blinding clarity of the conversional moment in the heart and not the mind. As the apostle Paul writes, "God, who commanded the light to shine out of darkness, hath shined in our hearts, to give the light of the knowledge of the glory of God in the face of Jesus Christ" (2 Corinthians 4:6).

The heart was also believed to be capable of regeneration. For medieval theologians, the human body had been miserably corrupted by Adam and Eve's fall from grace. This religious view of man found a grim confirmation in the physical world. Life on earth during this epoch saw the body deteriorate over the course of a lifetime at a rate that would shock most contemporary readers. The heart alone was deemed capable of regaining its lost innocence. Flesh might pucker from leprosy and organs die from untreatable ailments. But the heart of the believer could undergo a rebirth that was often seen as literal. Christ's healing power could actually restore the organ to a pre-Edenic purity. With the acceptance of the Savior into it, the stony, crumbling heart became supple and fertile. It became new. It is interesting to note how this thinking prefigures some of the most promising trends in contemporary cardiology, such as the use of stents to unblock calcified vessels and stem cell research that may soon allow the regrowth and replacement of heart tissue scarred by disease. For the early Christian, then, God was the supreme cardiologist. He opened passages clogged with sin, restored tissue scarred by vice, and returned the heart to its heavenly rhythm.

These multiple views of the heart as the repository of all that makes us human explain why, in an era when investigators had so little interest in delineating its physical function, people spent so much time cutting out hearts, scrutinizing them for signs of miracles, transporting them across daunting distances, and finding places to store them. The heart was the one indispensable corporeal item; it was the distilled essence of a believer. It could stand in for the whole man. During the Crusades, when it was impossible to ship the intact bodies of slain knights back to their European homes from the Levant—the eastern Mediterranean regions that were the Crusaders' targets—a warrior's remains were usually boiled down so that at least his bones might be transported. The heart was the only organ spared the cooking so it could be taken back whole. As was the case with Sister Chiara, the heart of anyone considered blessed was also liable to be excised and preserved. The dispersed burial of monarchs and other dignitaries was common. After Henry I's death in Normandy in 1135 from eating poisonous eels, his heart was sewn into the hide of a bull for preservation and transported back to England to be buried, while the rest of him was interred where it was. The heart of England's Richard I—whose nickname, Couer de Lion (Lionheart), is rumored to have come from his ripping out and consuming the heart of a lion to acquire its courage—had his legendary cardiac muscle buried separately from his other remains.

In addition to its overwhelming importance in the Christian's devotional mind-set, the heart also assumed a central role in secular discourse during the Middle Ages. The passionate source of humankind's deepest religious experiences was also seen, especially during the latter days of the epoch, as a wellspring of the romantic and erotic, often in ways that strongly contradicted the Catholic Church's moral teachings. One can almost feel the beating heart straining against these religious strictures. In poetry and the often bawdy songs of the troubadours, the heart became the vehicle for romantic love that could be clandestine, unrestrained, and, ultimately, damnable.

During this courtly era, when a knight's passion for his beloved lady was expected to remain chaste, the heart often supplanted the sexual organs as the bodily locus of sexual excitation and gratification. People made love with their hearts. Perhaps the most famous love story of the Middle Ages is that of Abelard and Heloise, the neutered monk and the young girl he was forbidden from seeing after their brief affair was so catastrophically discovered. Denied the opportunity for sexual intimacy, their hearts became the means by which they achieved consummation. Like nuns who would have ecstatic visions in which they exchanged their hearts for Christ's, Abelard and the student he had once seduced shared their most vital organs. "For my heart is not with me," Heloise writes climactically, "but with thee."

To lovers in the Middle Ages, the heart was the place in the body where their passion dwelled. It beat faster at the sight of the loved one and was seized with pain at a lover's absence or betrayal. In Giovanni Boccaccio's *The Decameron* (1353), the story of Prince Tancredi provides a particularly sanguinary take on the eroticized heart. In it, the aging and perhaps incestuously minded Tancredi catches his daughter, Ghismonda, making love with a commoner, Guiscardo. As punishment, the jealous prince has the man executed and eviscerated. His heart is then delivered to the young noble-woman in a chalice—Tancredi wants to prove that he not only has slain his daughter's lover, but also has severed the bond of love between the two rebellious young people.

Ghismonda, however, has an unexpected response. Instead of collapsing, she boldly cherishes this gruesome gift, "her tears gushing forth like water from a fountain [as] she implanted count-less kisses upon the lifeless heart." She then directly addresses Guiscardo's heart. "Oh, heart that I love so dearly, now that I have fully discharged my duties towards you, all that remains to be done is to bring my soul and unite it with yours." Tancredi's bloody spectacle has backfired. He has allowed the lovers to embrace once again, in a way that is perhaps even more intimate than their sexual congress. This reunion becomes eternal when Ghismonda adds poison to the concoction of tears and heart's blood, then drinks it down. As she dies, she places her lover's heart on her breast so that it might be close to her own heart as it stops beating. Ghismonda's ecstatic passion for the physical organ provides a worldly glimpse of the occasionally grisly devotions of her cloistered contemporaries. She has absolutely no doubt that Guiscardo's heart is sentient, capable of hearing his lover's endear-ing voice and feeling her gentle kisses—even after it has been so violently disembodied.

A similar eroticism is on display in Dante Alighieri's *La vita nuova* (1295). When the nine-year-old Dante first glimpses the love of his life, Beatrice, his reaction demonstrates the over-whelmingly central role that the heart played in medieval descriptions of erotic fascination and turmoil. With its evoca-tions of the three spirits that were the cornerstone of Galenic physiology, it also shows the continued hold that the great doctor's theories had a thousand years after his death:

> At that moment I say truly that the vital spirit, that which lives in the most secret chamber of the heart began to tremble so violently that I felt it fiercely in the least pulsation, and, trembling, it uttered these words: *"Ecce deus fortior me, qui veniens dominabitur michi*: Behold a god more powerful than I, who, coming, will rule over me." At that moment the animal spirit, that which lives in the high chamber to which all the spirits of the senses carry their perceptions, began to wonder deeply at it, and, speaking especially to the spirit of sight, spoke these words: *"Apparuit iam beatitudo vestra*: Now your blessedness appears." At that moment the natural spirit, that which lives in the part where our food is delivered, began to weep, and weeping said these words: *"Heu miser, quia frequenter impeditus ero deinceps!*: Oh misery, since I will often be troubled from now on!"

This primacy of the heart in processing the erotic can be seen again later, after the poet glimpses Beatrice on a crowded Florentine street and she greets him "so miraculously that I seemed to behold the entire range of possible bliss." The besotted Dante retreats to his room, where he falls into a deep sleep and has a dream in which he is visited by a lordly man who holds in his hand a flaming object. *"Vide cor tuum*," the man declares: "Behold your heart." And then Beatrice appears, only to be tricked by the man into eating the poet's smoldering heart, whereupon she ascends to heaven. Here, as in *The Decameron*, the heart is not simply a metaphor representing love but, rather, a sensory organ that a lover can caress and consume in order to meld with the beloved. Medieval physicians may have shied away from taking the heart in hand, from peeling it open and peering into its chambers, but the era's poets demonstrated no such queasiness.

By the time of Dante's death in 1321, the church's suffocating grip on science had begun to loosen. In medicine, the opening of trade routes to the East brought the work of Arab doctors to the West, and the fall of Constantinople in 1453 made available a flood of Greek texts that stimulated new interest in anatomy. Empirical inquiry was no longer equated with heresy. Some of the most talented minds of the era turned their focus away from heaven and back to the earth, especially to the long-neglected anatomy and physiology of the human being. The body once again became an object of curiosity. This change is vividly reflected in painted images, in which increasingly sophisticated materials and painterly techniques meant that the body no longer was presented as flat and austere; rather, it was radiantly beautiful and, often, unabashedly nude.

This new spirit of inquiry had a particularly powerful effect on the study of the heart. The Renaissance changed the way that humankind imagined the organ beating in the chest. Aided by the end of the taboo on dissecting human corpses, a group of singularly adventurous thinkers began to understand just how inadequate the theories of Aristotle and Galen were in explaining the heart's function. Elegant hypotheses would no longer suffice now that our hidden interior was finally revealed. God might still dwell in the heart's chambers, but that did not mean that man was incapable of taking their measure.

Chapter 3

Renaissance Heart

London, 1640

By the time he arrives in London, he has resolved to stop showing his wound. His journey through Europe exhausted him. He was supposed to be soaking up the continent's art and culture, but instead he found himself gawked at like a freak. And so, after establishing himself in rented quarters in Bedford Square, he starts doing something he has never before done— turning down invitations and requests. He refuses to attend salons and soirees; he sends back the calling cards of the doctors who arrive unannounced to examine him. If his wound cannot be cured, at least he will keep it hidden.

Then, just before he is to start his studies at Oxford, a request comes that he cannot deny. Sir William Harvey would like to examine him. The king's physician extraordinaire. By all reports the most brilliant man in the realm. Some say he is the greatest doctor since Hippocrates. What's more, Sir William has been ordered to make an account of business to King Charles himself, who has heard rumors of the wound. Turning him away is not possible. Hindering him in his mission would be treasonous.

And so Viscount Hugh Montgomery agrees to meet with the famous doctor. To let himself be poked and prodded and peered at like a corpse in an anatomy theater. Not that such indignities will be anything new. This has been his fate since he was thrown from his horse when he was ten—eight years ago, though the memory of the fall remains as fresh as if it had happened just last week.

He had been riding at his family's estate in Ireland when his horse stumbled on a rabbit hole. Tumbling to the ground, the young Lord Montgomery's chest had taken the full force of the impact against a jutting rock. There had been pain, of course, though at the time the worst thing about the incident was the breath being forced from his lungs. As he had sprawled gasping in the peat, he had thought that his end had come for sure.

The injuries had been grave. The ribs on the left side of his chest were badly fractured. The doctor, who had served in the army during the recent war with France, had blanched when he pulled back the boy's shirt to examine the wound. He'd announced that there would be no fixing the shattered bones. It was a miracle that they had not punctured the boy's lung or his heart. It was, in fact, a miracle that the young viscount was alive.

And then the fever had come, followed by the swelling. The skin over the sinkhole in his chest—a few inches below the collarbone and left of the sternum, where the rock had punched through—had quickly become raised and round and hard and hot to the touch. Pus had burst through, leaving a gaping wound that refused to heal. The hole was so deep that his throbbing left lung was visible. And it was big enough that three fingers and a thumb could be inserted. For months after that, doctors had gathered around his bed, some of them from as far away as Dublin—his father was a powerful man. A variety of cures were attempted:

poultices and plasters and cauterizing metals. Medicines that made him vomit blood and sleep all day and see demons dancing in the room's shadows. Nothing had worked, much to his father's growing disgust. The wound had remained stubbornly open. Finally his father had driven the doctors from the house and called in his blacksmith. The man's strong hands had trembled as he took measurements. But the small iron plate he forged fit snugly into the wound, held firmly in place by the remaining ribs.

The wound had never closed. But it had not killed Montgomery, either. He had learned to live with it as he grew into a man. What else could he do? It did not hurt; nor did it hinder his movements. He could not even feel it when visiting doctors touched the lung—or, rather, the thin, translucent leathery membrane covering the organ. Still, it had been no way for a boy to grow up, with a hole in his chest for everyone to stare at. For each day when he felt special, there were a hundred when he wished that he were the same as everyone else.

As Montgomery had grown older, the blacksmith had made new plates to fit his growing body. He could even ride again, though of course his father forbade him from allowing a horse to jump. The only treatment his wound required was a morning washing by his man, who would carefully lift off the plate and irrigate the exposed tissue with warm water from a syringe. Doctors from as far away as London and the Continent would visit to observe the motion of the naked lung. When they finally touched it, they always asked the same question: could he feel it? They always looked disappointed when he told them he could not. And then he always asked them the same question: could they fix him? His disappointment surpassed theirs when they told him no.

When he had turned eighteen, it had been decided that he was strong enough to tour Italy and France, as all young noblemen were expected to do before attending university. Everywhere he went, word had preceded him about the young man with a hole in his chest. In Padua and Paris he had obliged the native physicians, patiently enduring their prodding and questioning. He was invited to universities and hospitals; he was a guest in the homes of the rich and the curious. In Rome, there had even been a presentation at an opera house. It was sold out. Often ladies would swoon when they saw his wound. A few of the braver ones would ask to touch the pulsating lung, though most of them lost their nerve at the last moment.

The boldest of them all had been in Paris. A marquise. She was twenty years older than he, but also the most beautiful woman he had ever seen. She had weary eyes that looked as if they never again expected to be thrilled. But when he had removed his plate for her, those eyes had grown young and alive. She had not even asked permission before she touched the wound. Unlike the other women, who would poke at it shyly before dissolving into giggles or sobs, her fingers lingered against that thin, heaving membrane. She had stroked it as she would have a fine fabric. For the first time, he'd wished he *could* feel something there. And then she'd ordered her maid from the room and locked the door.

Since then, he had allowed no more demonstrations. Until Harvey.

The doctor arrives in the morning. He is a short, dark man with piercing eyes and a brusque manner. He wastes no time on pleasantries. He inspects the metal plate with a grunt of admiration and then orders it removed. Unlike other doctors, he shows no awe as he peers into Montgomery's gaping chest. He does not hesitate before inserting his fingers into the wound to feel the membrane and what lies beneath it.

He asks the usual question.

"Can you feel this?"

"No."

"Nothing?"

"Nothing at all."

He examines his patient for a while longer. Something appears to be perplexing him. Finally he looks up at the viscount, his small black eyes widening slightly in surprise.

"And they have told you this is your lung?" he asks.

Montgomery nods in confusion. Could this really be the king's physician? A man who cannot even recognize a lung when he sees it? Sir William once again inserts his fingers into the wound, although this time he also grips Montgomery's left wrist with his free hand. He appears to be counting.

"They are the same," he says as he finally lets go.

Montgomery does not understand.

"Sir?"

"Your chest and wrist pulsate at the same rate."

"What does that mean?"

"If this were your lung, the rates would be different."

"I do not . . . "

"It is your heart we are looking at." He points at the wound. "Your living, beating heart."

Montgomery cannot believe it. His first instinct is to presume that the man is a quack. But this is Sir William Harvey. So

Montgomery says nothing as the physician continues to examine his chest. The peevishness is gone now; something like wonder ripples through the deep lines on his face. It is as if he is seeing an old friend after a long absence. Montgomery remembers what he has heard about the doctor: The heart is his specialty. He understands it better than any man who has ever lived.

"We must go," he says eventually. "Immediately."

"Where?"

"Saint James."

It takes the viscount a moment to understand. They are going to see King Charles at the palace.

A rider is sent ahead to announce their visit. In the coach on the way over, Sir William is full of questions about the wound. But Montgomery has trouble focusing on the doctor's words. He is actually about to meet the king. He is not ready for this. When his father lived in London as a young man, he had met King James several times. But that was Montgomery's father, who feared nothing. He would often speak in rapt terms of the court's ceremony and majesty. One story in particular had stuck in Montgomery's mind. His father had been asked to join the royal entourage on a journey to some provincial cathedral, where James was greeted by subjects suffering from the "king's evil"— scrofula. One by one, these poor souls with their grossly swollen necks had approached to be touched by James. Although Montgomery's father was no believer in magic and superstition, he swore that several subjects had been cured on the spot. All signs of the disease that had so hideously disfigured the sufferer's neck had vanished. Kings truly do possess a healing touch.

The carriage finally arrives at the palace. Sir William is in such a hurry to get through the door that the equerries barely have time to spread the footcloth. The doctor is all business. It is almost as if he considers St. James's Palace to be *his* house. Montgomery's astonishment grows when he learns that they are going directly to the king's privy chamber—his private rooms. Sir William seems annoyed that the king is not waiting for them in the small, simply furnished room. He paces back and forth, muttering to himself, casting occasional dark glances at the young viscount as if he is in some way responsible for the delay.

A herald enters at last and instructs Montgomery on how to behave in the presence of the king. He must kneel when Charles enters. After he is told to rise he will remain standing unless instructed otherwise. He will speak only to answer direct questions, and then he must be as brief as possible. Wearing a hat is simply out of the question. And he must never, under any circumstances, attempt to touch the king.

After that, more waiting. Finally two guards enter, followed by several gentlemen who stare at Montgomery as if he is a cross between a Spanish assassin and a diseased cow. And then there is a scuffle of feet and everyone turns to the door. Montgomery feels as if his heart is about to leap through the iron plate covering it. A short man enters. He is so small that it takes Montgomery a moment to understand that this is the king of England. And yet, once that understanding comes, there is no mistaking his majesty. It radiates from him.

Upon his approach, Montgomery drops to his knees so quickly that he almost pitches forward. An annoyed Sir William motions for him to get back up. Charles does not appear to notice the viscount as he questions the doctor. The rumors are true—he speaks with a profound stutter. But there is intelligence and real curiosity in his conversation. And then his head swivels slowly, almost disdainfully, toward Montgomery. There is no greeting. The young man before him is simply a medical specimen. The king's eyes remain fixed on the viscount's chest as Sir William opens his blouse and removes the plate.

"In-ge-genious," the king says as Sir William holds up the device for him to examine.

Sir William impatiently beckons for Montgomery to step closer. After casting a worried look at the guards, the viscount does as commanded. The diminutive monarch needs to bend forward only slightly to put his eyes at the level of the young nobleman's chest.

"Your majesty will recall when I showed you the beating of a deer's heart?"

Sir William's tone is no longer curt. There is silk in it now. From deep in his coat, he produces a silver-tipped whalebone pointer.

"Notice that it is the same with a man," he says, gently jabbing the pointer into the wound. "Watch, if you would be so gracious, how in diastole—*here*—the heart's muscle draws in and retracts away from us. This is quite in contradiction to the motion of the heart itself, which is one of expansion."

The king nods. He is enthralled now.

"And then, and then—*here*—in systole, the heart comes forth and thrusts out, even though the muscle itself is contracting."

"There is such p-power in it."

"Systole makes the heart, Your Majesty. It is the action that sends the blood coursing on its long circuit to every district and borough of the body. And back again."

"Remarkable."

Sir William withdraws the pointer and stares at the king for a moment.

"Would Your Majesty condescend to feel its action?"

"Touch it?" the king asks, his astonishment chasing the stutter away.

"Yes."

"With my hand?"

"Yes."

King Charles thinks about this for a moment. And then a slight smile comes to his face.

"I sh-shall."

He pulls the pure white glove from his right hand—a herald scurries forward to collect it. The others in the room move a little closer. Montgomery can scarcely believe what is happening. The king of England is about to touch his heart. His astonishment is deepened by another thought. Perhaps he will be cured like the people his father witnessed at the cathedral had been. On the spot. His wound will close and he will no longer be a freak.

After a moment's hesitation, King Charles reaches his small hand forward and pokes his index finger into the wound. He has a slight tremor—arising from his power, Montgomery hopes. Surely he will feel the touch this time. The king is God's anointed leader. His healing hand is mightier than all the doctors in Europe combined.

But he feels nothing. Just the usual numb pressure, that

distant rumor of sensation. After a moment King Charles withdraws his finger. There is gentle applause around the room. As the king speaks again with Sir William, the young viscount waits for sensation to blossom around his heart, for the healing warmth to flood through his chest. He searches the eyes of the gawking courtiers for evidence of the wound's miraculous closure. But everyone's attention is focused on the king.

Montgomery understands. There will be no miracle. There is no healing power in the king's touch, at least not for him. King Charles is just another curious man. Sir William continues to speak, once again waving the whalebone pointer over the wound as he lectures the king on the heart. Montgomery does not even try to follow what is being said. His disappointment is too acute. All he can think about is having this crevice in his chest for the remainder of his days.

Finally the exhibition ends. The king sweeps from the room like a sudden breeze, without so much as a nod in the direction of the man whose heart he has just touched. The courtiers follow him without looking back. Sir William turns to Montgomery, expecting him to share in the wonder and majesty of the moment. But his keen eyes immediately detect that there is something bothering the younger man.

"Are you in pain?"

"No. I just thought . . . "

"What?"

"That there would be a miracle," he confesses. "That his touch would heal my wound."

Sir William frowns.

"Your wound will never close, I'm afraid. Only an operation

could accomplish that, and it would be too dangerous at this point. Your body has adapted. As bodies do. You will be like this until the day you die."

Montgomery nods. Sir William continues to watch him. There is a fierce sympathy in those small black eyes.

"Come with me," he says finally.

He takes the viscount by the sleeve and leads him over to a large mirror. He pulls back some thick curtains at an adjacent window so that there will be enough light. As the doctor takes up a position next to him, Montgomery peers into the wound. It has been a long time since he looked inside himself. It seems different now. A lung is just . . . a lung. But this is his heart. The very essence of him. What the king had said was true. There is such power in it! It seems to want to leap right out of his chest. He cannot understand how it manages to move so regularly, so forcefully, without his ever commanding it.

"*This* is the miracle, if a miracle is what you are after," Sir William says. "I have been studying the heart for thirty years now, and I never cease to be in awe of its power and its grace. Throughout my whole career, I have seen people looking for miracles. Witchcraft and holy relics and, yes, the touch of a king. And all the while, it is right here inside them."

Montgomery nods. He knows that the doctor is right. But there is no real comfort in that knowledge.

"Cheer up!" Sir William says. "Most men strive after fame their whole lives. You've achieved it simply by tumbling from a horse."

Montgomery finally manages a thin smile.

"Now, come with me to St. Bartholomew's Hospital, young man. I would like to show you off to my colleagues."

"Of course," Montgomery answers, taking one last look at his heart before the physician covers it with iron.

IT started with a crime. The revival of interest in the heart's anatomy can be traced to Italy in 1315, when an anonymous woman was sentenced to death for a now-forgotten offense. Soon after her execution, a Bolognese professor and surgeon named Mondino de' Luzzi took possession of her corpse and performed the first known legal human autopsy in the West in a thousand years. The Vatican authorized the procedure, deeming it an acceptable extension of the sinning woman's punishment. After all, the fate that awaited her in the afterlife was far worse than the indignity of being sliced open in a lecture hall.

De' Luzzi's scrutiny of that dissected corpse was not particularly enlightening, since he did little more than use it to confirm Galen's erroneous theories. On a more profound level, however, the professor's bold expedition beneath the skin was to prove every bit as momentous as Columbus's voyage into another unknown territory at the end of the following century. Like Columbus, De' Luzzi did not reach his intended destination or comprehend where he had actually arrived. But he did pave the way for others who would understand this new terrain. The long journey to open heart surgery, balloon catheterization, and magnetic resonance imaging technology—the latter of which can now map every millimeter of the cardiac muscle—had begun.

With the advent of human dissection, physicians no longer had to peer into the entrails of slaughtered animals in an attempt to understand the heart; they no longer had to speculate about its form and function. From the moment de' Luzzi opened up that forgotten woman's thorax, Galenic theory was doomed. Cutting up dead bodies would make the heart come alive in unexpected ways. Dissection would demystify the muscle's operation, but it would also deepen humankind's awe of its power and complexity. The heart was finally able to speak for itself. And the tale it told was more astonishing than Aristotle or Galen had ever guessed it would be.

Although dissection would eventually chase God from the physical heart, the initial motivation to allow these procedures was religious. As Jonathan Sawday points out in his authoritative study of the subject, *The Body Emblazoned: Dissection and the Human Body in Renaissance Culture*, early Renaissance scientists were not looking for something new when they opened the chests of executed criminals. Rather, they were seeking to confirm beliefs that were very old. Anatomies were intended to reveal not just evidence of God's handiwork but also his actual presence. These men were digging for God.

It is hardly surprising, therefore, that these early proceedings had more in common with religious ceremonies than with modern autopsies. Usually performed in a circular, candlelit theater, they would begin with the body being raised from below via a trapdoor. A hush would fall over the audience, which was normally made up of students, noblemen, and thrill seekers. People might have grown accustomed to the sight of dead bodies during this era of plague and war, but this was different. The anatomist, dressed in a robe and cap, would then enter and perch above the corpse in a chair or at a lectern. As he began to read from a text—Galen, or perhaps Aristotle—his assistant, usually a barber who worked as a surgeon on the side, diligently sliced and peeled and exposed the corpse to provide graphic illustrations of the master's words.

These presentations were especially beholden to tradition when the anatomist was describing the heart. Nonexistent features of the organ, such as a third ventricle or pores in the septum, were confidently detailed, even though closer scrutiny would have revealed that they simply did not exist. It was almost as if the learned professor refused to believe what his eyes were seeing.

A very different motivation for human dissection came with the painters and sculptors of the fifteenth and sixteenth centuries. With the era's new humanist focus, Greek and Roman myths joined Christian stories as acceptable subject matter for artwork. This brought about a return to classical imagery, which often focused on the nude figure. The human body was no longer shrouded in shame. Increasingly sophisticated artistic techniques in rendering perspective, shading, and dimension also inspired artists to explore methods of representing the body as accurately as possible. To do this, an understanding of its inner scaffolding of muscle, vein, and bone was required. Some painters would become as adept as physicians at anatomy. The era's premier writer on the visual arts, Giorgio Vasari, claimed that the fifteenth-century sculptor Antonio Pollaiuolo was the "first master to skin many human bodies in order to investigate the muscles and understand the nude in a more modern way." Particularly striking were the era's many drawings of partially flayed limbs, which showed the subject's skin literally peeled back to reveal the muscle and bone beneath. Physicians may have been hindered by their allegiance to their Greek and

Roman predecessors, but the hunger of Renaissance artists to work in a classical tradition proved liberating.

The most famous of these artist-anatomists was Leonardo da Vinci, whose notebooks contain hundreds of drawings of the human body, including several highly detailed illustrations of the human heart and vascular system. Legend has it that da Vinci worked with corpses obtained by methods that were not always strictly legal. He was no mere illustrator, however. He may initially have viewed the cardiac muscle with an artist's eye, but he soon brought a scientist's curiosity to bear on the subject. His theories on the heart, rendered in his notorious mirror writing, could be breathtakingly sophisticated. They were also, on occasion, naively indebted to Galen and Aristotle. For instance, like the Greeks, he thought the heart's main purpose was not to circulate blood, but rather to create heat, which it did through the friction caused by the movement of blood back and forth between its chambers. Da Vinci proved far more astute when it came to the function of the heart's valves. In a series of brilliant sketches, he demonstrated how the tricuspid valve's three leaflets are sealed by whirlpools caused by the blood that has just passed through them. His drawings of the coronary arteries are also remarkably realistic—a modern reader riffling through a book of da Vinci reproductions could be forgiven for thinking he was instead looking at CT scans of human hearts.

The scientific and the artistic became fully integrated with the work of the great Renaissance physician Andreas Vesalius, whose insights are contained in the greatest anatomical book of its time, the richly illustrated *De Humani Corporis Fabrica* (*On the Fabric of the Human Body*) of 1543. Vesalius's breakthrough came when he discarded Galen's erroneous belief that venous blood passed through tiny pores in the heart's septum, moving from the right ventricle to the left to be mixed with air inhaled from the lungs. After repeatedly dissecting the hearts of both animals and human beings, Vesalius concluded that these small holes simply did not exist. There had to be another way for the blood to move from the right side of the heart to the left. And there had to be another reason the vivid scarlet blood in arteries looked so different from the darker maroon stuff running through the veins.

His colleague Realdo Colombo came up with the answer: pulmonary circulation. (In fact, this breakthrough may actually have been made by a Spaniard, Michael Servetus, but his manuscripts were used as kindling when he was burned at the stake in Geneva on unrelated charges of blasphemy.) During vivisections of dogs, Colombo saw that the pulmonary veins, four vessels that travel from the lungs to the left side of the heart, were filled with only bright red arterial blood. He knew, then, that Galen's model could not be correct. Blood was somehow being turned bright red *before* it reached the left ventricle. With this in mind, Colombo proposed a new model in which blood traveled from one side of the heart to the other in a circuit. Starting on the right side, it moved to the lungs, where it was mixed with air to become, in his words, "shining" and "beautiful." From there it was returned to the left side through the pulmonary veins before moving on to the body at large.

With this elegant theory, a new idea had been introduced, one that suggested a major break with everything previously believed about the heart's function and the movement of blood through the body. Blood *circulated*. Its course through the body was not always a one-way street in which it was used up at the end of the journey. The vast system of veins and arteries that thinkers had been puzzling over for millennia looked to be even more complicated than they had suspected. And yet, while Colombo's model marked a great advance, the picture it painted was still far from complete. Although his vision of pulmonary circulation was basically accurate, he still held to the traditional view of the blood's movement through the body. The big picture was still lacking. This would have to wait another seventy years, until a slight, temperamental Englishman—a devout Christian and the personal physician to two kings—changed the way we think about the heart as no one has before or since.

LEGEND has it that the smallest incidents can often lead to the greatest scientific breakthroughs. Newton's apple, Galileo's rock versus feather—even if these stories are not strictly true, they help the layman understand mental leaps that might otherwise be imponderable. In the case of Sir William Harvey, his moment of clarity came when he pressed a finger against a forearm. From this simplest of acts, he began to completely reimagine the way that we think about the heart. Its function became all the more transparent—and all the more awe-inspiring.

While Harvey's name is not as familiar as Newton's or Galileo's, his contribution to humankind's understanding of the natural world is every bit as profound. Born in 1578, he studied at Cambridge and then in Italy under the great surgeon Hieronymus Fabricius. After returning home, he served as personal physician to James I and later to James's son Charles I. In 1628 he published a short book with a long title: *Exercitatio Anatomica de Motu Cordis et Sanguinis in Animalibus* (*Anatomical Exercise on the Motion of the Heart and Blood in Animals*). With this elegantly written document, known now as simply *De Motu*, the errors of Aristotle and Galen were finally laid to rest. Though he was personally conservative, Harvey set forth one of the most revolutionary agendas in the history of science when he announced in the Introduction that "it is plain that what has heretofore been said concerning the movement and function of the heart and arteries must appear obscure, inconsistent, or even impossible to him who carefully considers the entire subject." Combining a remarkably acute clinical sensibility with a strong dose of British common sense, he provided a paradigm for the heart's function every bit as transformative as Galileo's view of the movement of the earth.

What Harvey discovered is so commonplace for the contemporary reader that it is hard to imagine that people could ever have thought otherwise: blood travels in a continuous circuit through the human body. There is only one vascular system, not the two networks that Galen had posited. And blood does not move because it is attracted by hungry organs or hustled along by quivering vessels, as was previously thought. Its movement is all about the heart, whose powerful systolic contraction sends blood out on a long, life-sustaining circuit that reaches every precinct of the human body.

Harvey came to this insight through a series of remarkable experiments and calculations. The first showed that the veins carried blood *toward* the heart, not away from it, as Galen and his followers had thought. Harvey determined this by applying a tourniquet and isolating and applying pressure at two points along a vein on his forearm, then pushing his finger along its course toward the elbow. The vein's contents moved easily, causing the vessel to bulge in front of the finger. When he tried to push it down the arm toward the wrist, the blood appeared to resist being hustled along, and the vein seemed to vanish into the muscle. This same procedure worked in the leg—venous blood moved easily toward the heart, but struggled when forced away. And when he ran his finger up and down the neck's veins, blood could easily be forced toward the chest, but not upward to the head.

From this, Harvey was able to conclude that veins, wherever they may be located in the body, carry blood in the direction of the heart. He confirmed this by tying ligatures slightly above venous valves of vivisected animals and then clearing the isolated portion of the vein of blood by pressing. When he did so, the isolated section of the vein downstream, or away from the heart, remained empty. Upstream, or on the heart's side of the isolated section, the blood pooled between the valve and the ligature, bulging the vessel. This established that valves in the veins did not prevent blood from swamping the extremities, as had previously been thought, but rather kept it from backsliding on its journey toward the heart.

Harvey also completely changed the way we think about the heartbeat. He determined that it was a two-part action. "There are, as it were," he wrote in *De Motu*, "two motions going on together: one of the auricles [now commonly called the atria], another of the ventricles; these by no means taking place simultaneously, but the movement of the auricles preceding, that of the heart following; the movement appearing to begin from the auricles and to extend to the ventricles." Although physicians before Harvey understood that the dilation and contraction of the heart were involved in the movement of blood, it was thought that the key part of this double action was the heart's expansion. As we explained in Chapter 1, Galen and his followers thought the dilated heart acted like a bellows that drew in air to cool its innate heat. Through close observation of vivisected animals, Harvey determined that the opposite was true. It was the heart's contraction—its systole—that did the heavy lifting to propel the blood through the body.

To picture this, he used two very contemporary metaphors. The first compared the heart to a flintlock rifle, in which the fall of the hammer sets off a series of events that lead to the expulsion of the bullet "in the twinkling of an eye." In the heart, the contraction of the atrium was like the fall of the rifle's hammer, forcing blood into the ventricle, from where a second, more powerful detonation would fire it into the body.

The second mental image Harvey employed compared the heart to a pump. As the author and physician Jonathan Miller astutely pointed out in *The Body in Question*, Harvey was greatly aided here by living in an era when pumps were first being widely used in mining, industry, firefighting, and agriculture. Humankind has, after all, always employed metaphors to think about the heart. The Egyptians conceived of it as being like the source of the Nile, Aristotle as the body's furnace, and the early Catholic thinkers as the earthly dwelling place of God. Harvey's genius was to use, for the first time, a metaphor that accurately conveyed how the heart actually worked.

His most important experiment involved a quantitative calculation. If Galen was right and blood flowed only outward from the liver and heart, Harvey reasoned that the amount of blood sent forth would have to be equivalent to the amount consumed by the muscles, organs, and bones that it nourished. According to Galen, it would all be used up. After reckoning the volume of the heart's chambers and their efficiency in pumping blood, Harvey could see that there was no way the amount of blood being issued under the Galenic model could possibly be consumed by the human body. A person would drown in a matter of hours. The blood had to go somewhere else. The only theory that made sense was for it to return to the heart. Harvey bolstered this argument by noting that drugs and poisons acted on the entire body, often with startling rapidity, no matter where they were injected. One interesting result of this breakthrough was that it undermined a key rationale for the long-standing practice of bloodletting. If blood flowed constantly throughout the entire body, then removing it from one particular area by opening a vein or applying leeches made no sense.

Although Harvey's discovery of circulation formed the basis of how we now think about the heart, his views were in several important respects incomplete. Although he deduced that there was some invisible means of transferring blood from the outgoing arterial to the incoming venous system, it was not until 1660 that the Italian physiologist Marcello Malpighi, a pioneer in the use of a newly minted device called the microscope, actually saw blood move through capillaries. Nor did Harvey understand that the reason blood circulates through the lungs is to become oxygenated— this was discovered by Richard Lower in 1669. Instead, Harvey maintained that blood possessed its own innate "vital forces," which it transported to the body in order to keep it alive. The ghosts of Aristotle and Galen had yet to be completely exorcised.

For all of his reliance on rigorously empirical approaches to studying the heart, Harvey ultimately imagined the organ in spiritual terms. It might be a pump, but God was working the handle. Perhaps influenced by his service as physician to the monarchs James and Charles, Harvey deployed another metaphor for the heart: it was the body's king, its anointed, the internal manifestation of a universal order. In this regard, at times he seemed to believe the heart to be as deeply infused with God as his medieval forebears had. This was evident from his famous first lecture to the Royal College of Physicians of London, which began with the following lines, in Latin, from Virgil: "I begin with Jove, O muses. All things are full of Jove." (Both Harvey and his audience would have immediately substituted the Judeo-Christian God for the pagan Jove.) Harvey may have explained the heart's function in terms of a gun or a pump, but in its essence it was much more than a machine. It stood at the center of God's blueprint for ordering the world. It was a notion that Harvey expressed with typical elegance in dedicating *De Motu* to Charles I:

> Most serene King! The animal's heart is the basis of its life, its chief member, the sun of its microcosm; on the heart all its activity depends, from the heart all its liveliness and strength arise. Equally is the king the basis of his kingdoms, the sun of his microcosm, the heart of the state; from him all power arises and all grace stems.

But Harvey's religious devotion could not stop the empirical revolution he had started. In *Discourse on the Method of Rightly Conducting the Reason and Seeking the Truth in the Sciences* (1637), published less than a decade after *De Motu*, the French philosopher René Descartes gave Harvey's pump metaphor a radically materialistic expression. For Descartes the body was a mechanism, pure and simple. It was a "mere mechanical contrivance," an automaton whose movements followed the same principles as a clock. It had moving parts; it suffered breakdowns; it required motivating fuel; and, if worked on by skillful-enough hands, it could be repaired. It was kept alive by the same force that animated a cat or a begonia.

The thinking ego, the "I," whether it was called soul or mind, was distinct from the body. Although Descartes accepted the existence of a soul, he confined it to the brain's pineal gland and said it was capable of willing only certain actions. It was not responsible for the most basic bodily functions—especially the heartbeat. Mind and heart were radically bifurcated. Descartes scorned the idea that blood contained some sort of vital spirit that made life possible. For the first time, a major Western thinker was able to imagine the heart without recourse to concepts of God or the soul. The ghost had left the machine. For the next century and a half, through the period known as the Enlightenment, Descartes's radical theories would liberate scientists to begin fledgling inquiries into this glorious clock's ticking, investigations that would lay the groundwork for the remarkable breakthroughs of the past two hundred years.

ALTHOUGH Harvey published *De Motu* in 1628, notes from his Lumleian Lectures to the Royal College of Physicians of London suggest that he first presented his views on the basics of human circulation in 1616, in a talk given just one week before the death of William Shakespeare. While these two singular geniuses were nearly contemporaries, their influence on our understanding of the heart was in many ways antithetical. Even as the era's premier physician set humankind on a course to understand the physical heart in purely mechanistic terms, its greatest poet was able to inoculate the heart's symbolic power against infection by the empirical. In doing so, he would keep alive the long tradition that saw the heart as the source of our emotions and, ultimately, of our identity. After Shakespeare, nothing science could do would be able to deny the heart its metaphorical significance.

The word *heart* appears more than one thousand times in Shakespeare's plays and sonnets. *Love* merits only a handful more mentions. Certainly, the Shakespearean heart incorporates most of the qualities it had previously been associated with, such as courage and erotic desire and fidelity. But the heart has another meaning for Shakespeare—one that may not be entirely unique to him but that certainly takes on a new profundity in his work: it becomes clandestine. For Shakespeare, more than any writer who preceded him, the heart becomes an individual's hiding place, the secret cellar where thoughts and feelings can be spirited away.

The heart is a vital piece of stagecraft in Shakespeare's plays. It often serves as a costume or a prop. It can be flourished like a pure white robe or hidden like a cloaked dagger. Shakespeare's characters are often defined by their willingness to reveal the contents of their hearts. Villains hang a thick veil between the heart and the world; their words and actions often misrepresent their closely held thoughts and feelings. Worthy characters are reluctant, and sometimes constitutionally unable, to sever their spoken words and overt acts from their hearts' impulses.

The virtuous, bared heart is best represented by *King Lear*'s Cordelia (whose name can be interpreted to mean pure or warm of heart). Lear's daughter has no veil between what she speaks and what she genuinely feels. She wears her heart on her sleeve. When her father demands that she describe her love for him in terms as fulsome as those used by her duplicitous sisters, Cordelia protests that she is incapable of this kind of emotional extravagance: "I cannot heave / My heart into my mouth." She does not mean here that she cannot speak what is in her heart. Quite the opposite—she will not transplant her heart from its natural position at the core of her being into her mouth, where empty words are often minted. Furthermore, she explains that once she is married, she will be able to give her father only half of her love. The rest of her devotion will go to her husband.

Mistaking his daughter's honest heart for a cold one, Lear explodes with rage. "But goes thy heart with this?" he demands. Audience members, even though they have known the princess only a few minutes before this exchange occurs, will have no trouble answering the question in the affirmative. Cordelia is clearly the type of person whose heart is in everything she says. Only a fool would think otherwise. But Lear, as his own Fool will often remind him, is precisely that.

What feels new in the play is Shakespeare's suggestion that sincerely speaking one's heart is no longer sufficient to win the day. Although Christian martyrs were executed for being true to their hearts, they would get their rewards in heaven. No such outcome is assured Cordelia and her father. Their deaths have a distinctly nihilistic air about them. The sincere heart's role in the world is suddenly challenged. The old order, in which an honest heart stands at the center of a just universe, might not have been toppled, but it is certainly given a good shaking. Lear, the king, is the heart of the play's body politic, and yet he is more interested in words that flatter his ego than in those that express inner truths or confirm universal ones. In Shakespeare, the heart may still be home to an individual's deepest self and the fountainhead of strong feelings, but this does not mean that unlocking its doors and throwing open its windows is a beneficial strategy. Cordelia's purity is rewarded with rejection, exile, and slaughter; Lear's belated understanding of the contents of her heart (and his own) cannot save him from a similar fate. His daughter's death is accompanied not by a heavenly choir singing her to her reward, but by the howls of a heartbroken old man who understands his blindness too late.

Shakespeare's villains also embody a new way of thinking about the heart. A wicked heart is no longer necessarily punished, at least not any more strenuously than a noble one. Shakespeare's rogues were a varied crew, some of them out for political gain, others simply wreaking havoc to quench what the English poet Samuel Taylor Coleridge called their motiveless malignity. The one quality they all share is the ability to mask their intentions, which often spring from the heart with an intensity and randomness that surprises even the villain.

This makes for a new kind of rogue, one who might not choose to be evil, but who certainly prefers to keep his black heart under wraps. Take *Othello*'s Iago, who swears to keep "his heart attending" to only his own interests, even as he outwardly pledges devotion to Othello, the master he secretly reviles. As the play opens, Iago sets out his agenda in terms that perfectly express Shakespeare's view of the clandestine heart:

> In following him, I follow but myself;
>> Heaven is my judge, not I for love and duty,
>> But seeming so, for my peculiar end;
>> For when my outward action doth demonstrate
>> The native act and figure of my heart
>> In complement extern, 'tis not long after
>> But I will wear my heart upon my sleeve
>> For daws to pick at: I am not what I am.
>> (1.1.58–65)

Put differently, only a fool would wear his heart on his sleeve for others to tear apart. That risks the sort of destruction meted out to Cordelia. The cunning man fabricates an artificial heart to keep on display.

In Shakespeare, the heart's deepest truths are not necessarily divine. The heart is no longer imagined as solely a vehicle for the transmission of God's word. And evil is no longer depicted as simply ignoring holy commands. Unmoored from its scriptural anchor, the heart becomes unruly and unpredictable. The thoughts and feelings it generates, good or evil, can destroy its owner. Lovers who answer its call are as likely to wind up punished as they are rewarded; corrupt villains can ride roughshod over their innocent peers, doing untold damage before the true nature of their black hearts is uncovered.

The Shakespearean heart is confounding. Time and again,

characters express surprise at the passions bubbling up from this volcanic source. Angelo, in *Measure for Measure*, is a fundamentally good man who is driven to malicious acts by his overwhelming attraction to Isabella. He seeks guidance from above, though he acknowledges that in his "heart the strong and swelling evil" of his lust for her voids his prayers. Word of her approach has a disabling impact on him:

> Why does my blood thus muster to my heart,
>> Making both it unable for itself,
>> And dispossessing all my other parts
>> Of necessary fitness?
> (2.4.20–23)

Here the heart is pictured as startling, disruptive, greedy, selfish, and irresistible. Love is a sort of cardiac shock. The heart is no longer necessarily an ally but, rather, a potentially mutinous force inside the individual. It is capable of great loyalty and service, though it is also liable to lead an uprising that the mind and the will are powerless to put down. As the French philosopher Blaise Pascal would say several decades later, "The heart has reasons that reason does not know."

It is in *Antony and Cleopatra* that Shakespeare's use of the metaphorical heart reaches its fullest expression. Traditional conceptions of the heart are merged with new imaginings to create one of the most heart-rich works of art in existence, one that can serve as a guide to many of the metaphorical uses of the heart before and since. The play's two charismatic lovers possess very different hearts. Antony's is that of a conqueror. It is

Greco-Roman. Muscular and armored, it houses his strength and courage. Cleopatra's heart is Eastern. It is intuitive, mercurial, as supple and textured as rumpled bed linens. There is something of the goddess about the queen—her heart is allied with the eternal. As reported by Antony's sardonic friend Enobarbus, who is anything but starry-eyed, it would be a mistake to "call her winds and waters sighs and tears; they are greater storms and tempests than almanacs can report."

The mixture of these two hearts proves combustible. The play opens with a Roman officer complaining that Antony has surrendered his heart—one of the most fearsome weapons in the Roman arsenal—to the Egyptian queen.

> His captain's heart,
>> Which in the scuffles of great fights hath burst
>> The buckles on his breast, reneges all temper,
>> And is become the bellows and the fan
>> To cool a gipsy's lust.
>> (1.1.6–9)

The metaphor of a mutinous heart takes on an almost literal meaning here. By giving his heart over to Cleopatra, Antony has denied his essential valor and betrayed his fellow soldiers. His capitulation cannot withstand the pressures of geopolitics. His coruler, Octavius Caesar, needs Antony's warrior heart in order to quell a very real mutiny in their empire. But during a key naval battle, Antony displays cowardice—faintheartedness—by following his lover when she retreats. Cleopatra's heart, after all, was made not for military combat, but for indoor pursuits.

Although Antony tries desperately to regain his honor in subsequent battles, it is too late. He has allowed Cleopatra to be the "armourer of [his] heart," and her shield cannot protect Antony from Caesar's iron—or his own sense of shame. After believing a false report of her death, the great captain realizes that his dream of merging his heart with Cleopatra's is now impossible. The only way to salvage honor is to plunge his long-neglected sword into his chest to cut free his captive heart.

It is after his death that Cleopatra displays the true quality of her own heart. In the play's early acts, she treats her heart as just another jewel that adorns this most beautiful of queens. There is something theatrical about the way she uses it. Once her lover dies, however, her heart undergoes an apotheosis. Although she was a coward in battle, she proves a hero in love, holding the poisonous asp to her breast so that its venom can quench her "immortal longings" to join Antony, whom she unforgettably pictures dwelling in eternity. As the venom flows through her veins, her liberated heart, in the classical Egyptian formulation, flows into the eternal Nile and the starry heavens. "Husband, I come! / Now to that name my courage prove my title! / I am fire and air." (5.2.286–288) In death, the Roman and Egyptian hearts finally become one. The heart's mutinies are quelled, and cowardice and artifice are banished. Courage and honor cohabit with instinct and erotic passion. At this moment, Shakespeare reconciles two apparently conflicting views of the heart, and the Renaissance heart finds its most complete artistic representation.

With Shakespeare, the heart's metaphorical identity underwent a shift that allowed it to survive the ongoing assault by science. The heart had been laid bare, plucked from within the ribs, and sliced open. No god was discovered inside. And yet the heart endured as an image for the forces, good or ill, that could arise unbidden in the individual. The heart still leapt at the sight of a loved one who perhaps should not be loved; it still raced in the heat of a seemingly lost battle, giving the warrior the strength he needed to overcome impossible odds. Something was going on in there, something deeper than thought. By becoming a metaphor for incomprehensible passions and hidden selves, the Shakespearean heart staked a claim to ground that the microscope and the scalpel could not violate.

NOT all artists were able to resist the lure of using the scientific revolution as a reason to sap the heart of its sublimity and turn it into a simple plot mechanism, just as their peers in anatomy were turning it into an engine in the human machine. This approach was particularly pronounced in the work of the generation of Jacobean playwrights who followed Shakespeare, most notably John Ford. Where Shakespeare exploited and expanded the vast inventory of the metaphorical possibilities of the mutinous heart, Ford was more concerned with exploring its practical utility in the dynamics of his turbocharged plots.

This is readily apparent in the first of his two great tragedies, *The Broken Heart*, a tale of forced marriage and bloody revenge. Its last two acts include the sort of carnage endemic to Jacobean drama. After one character is knifed while trapped in a trick chair and another starves herself to death, a third, Orgilus, is condemned to death by a means of his own choosing. He opts "to bleed to death" under his own command, eschewing the use of a professional surgeon. In this he is assisted with unseemly glee by Bassanes, his rival for the affections of the recently deceased anorexic, Penthea. When asked whom he will have as his executioner, Orgilus responds:

> ORGILUS: Myself; no surgeon.
> I am well skilled in letting blood. Bind fast
> This arm, that so the pipes may from their conduits
> Convey a full stream. Here's a skilful instrument.
> [Shows his dagger]
> Only I am a beggar to some charity
> To speed me in this execution,
> By lending thither prick to th'other arm,
> When this is bubbling life out.
>
>
> BASSANES: I am for 'ee.
> It most concerns my art, my care, my credit.
> Quick, fillet both these arms.
> (5.2)

The language here is coldhearted and grimly efficient. Veins are "pipes" and "conduits"; blood "bubbles" from a filleted arm. The subtlety of characters such as Lear and Cleopatra has given way to the language of the civil engineer and the butcher. The Cartesian view of the heart as a fleshy machine predominates. The play's final death—the collapse of the beleaguered princess Calantha from a broken heart—seems almost quaint in comparison to the preceding gore. Looking at her unblemished, white-clad corpse, one might be tempted to think that she got off lightly.

Ford's *'Tis Pity She's a Whore*, the most notorious drama of the Jacobean era, further objectifies the heart. There is no doubt about the intensity of the incestuous passion between Giovanni and his sister Annabella—early on, he declares his desire that they constitute "One soul, one flesh, one love, one heart, one all." Again, however, the heart is imagined less as a source of love than as a prop to be torn, stabbed, pulverized, immolated, bartered, or displayed. When Giovanni declares his sexual desire for Annabella, he offers his heart not as a fond token but rather as a target for his dagger, should she take offense.

> And here's my breast, strike home.
> > Rip up my bosom; there thou shalt behold
> > A heart, in which is writ the truth I speak.
> > (1.2)

When Annabella's suitor Soranzo deploys traditional romantic terminology to tell her that he is "sick to th'heart" with love for her, she cruelly mocks him by offering to fetch him some medicine. Nobody can use heart conceits in Ford's drama, it seems, unless he or she implies a trip to the emergency room. This transformation of the heart into a blood fetish becomes complete in the play's notorious climax, in which Giovanni transports his murdered sister's heart into a banquet on his dagger's tip.

I came to feast, too, but I digged for food
>In a much richer mine than gold or stone
>Of any value balanced; 'tis a heart,
>A heart, my lords, in which mine is entombed.
>(5.6)

Informed by the great Italian anatomists as well as by Harvey's image of a blood-pumping mechanism, Ford and his contemporaries turned the heart into the main course, to be served rare at their bloody dramatic feasts.

AT the same time that anatomists were making the heart visible and Harvey was delineating its function, the Protestant Reformation was bringing about another revolution in the way the heart was imagined. With the proliferation of anatomy theaters and increasingly accurate illustrations of the cardiac muscle, the medieval belief that the heart could actually house God or his relics was no longer tenable. Investigators had combed the scene, and God's fingerprints were absent from the myocardial tissue. A new method of accommodating him in the heart's chambers had to be developed, one that abandoned literal iconography for imagery more in keeping with the tenor of the time.

And it was not just the pressures of the anatomical revolution that forced a new way of imagining the Christian heart. The radical theology of the Protestant Reformation also demanded that it change. Although John Calvin may have delayed the breakthroughs of the anatomists by a few years when he ordered that one of their fraternity, Michael Servetus, be burned at the stake for heresy, he also played a key role in changing the way we think about the heart. The most influential of the Reformation thinkers, he demanded that believers turn away from what he saw as the idolatry of the Catholic sacrament. In his central work, *Institutes of the Christian Religion*, Calvin thundered that "every thing respecting God which is learned from images is futile and false." To Calvin, God did not need an icon; the Logos—divine wisdom—could do without a logo. The only images that Calvin deemed appropriate were those that "gave plain intimation of his incomprehensible essence. For the cloud, and smoke, and flame, though they were symbols of heavenly glory (Deut. 4:11), curbed men's minds as with a bridle, and that they might not attempt to penetrate farther."

Given these interdictions, to Calvin, the Catholic Church's ongoing use of the heart as an emblem of Jesus Christ was not only erroneous, but also sinful. Calvin's outrage was piqued by the ever-more-vivid representations of Christ's heart by Renaissance painters, who used their newfound techniques to create an organ worthy of idolatry. In Calvin's mind, such an image was not much better than a carved totem worshipped by savages. In addition, a group of nuns adoring a crucifix lodged in a human heart would have been not only anathema for him, it was the sort of idolatry that had helped provoke the Protestant Reformation in the first place. The same held true for the Eucharist. The Catholic view that communion wine actually *became* Christ's blood was yet more idolatry. Calvin preferred a far more bloodless way of picturing the sacrament, claiming that the wine was infused with a "pneumatic presence" of the Holy Spirit.

Iconoclasm such as Calvin's resulted in a number of eruptions of violence against churches and the artwork contained within them. In the Beeldenstorm ("The Smashing of Icons") in Holland and Belgium in 1566, thousands of works of art were destroyed by mobs whipped into a frenzy by Protestant preachers. These would undoubtedly have included many representations of Jesus's and Mary's hearts. During the English Civil War in the following century, Oliver Cromwell's government gave Puritan fanatics such as William "Smasher" Dowsing the explicit task of destroying idolatrous art, leaving a trail of broken hearts behind them.

This did not mean that there was no place for the heart in Protestant worship. In fact, its symbolic potency remained strong. Its significance was simply relocated. Under Calvin's influence, the heart was envisioned exclusively as an instrument of understanding, a sort of deeper mind that allowed the believer to comprehend the meaning behind the smoke, clouds, and pneumatic presence. For the Protestant thinker, the only way to attain knowledge of God was through studying the scriptures. The mind simply was not up to that task. It might comprehend the surface meaning of scriptural text, but it could never grasp its essence. Because the Catholic Church's authority in these matters had been rejected, Calvin and his followers needed another means of understanding the word of God. This is where the heart came in. It alone could truly *know* the Lord. As Protestant doctrine developed, the heart came to be viewed as a sort of personal translator for decoding and understanding the books of the Bible.

This spiritual acuity was not something that could be gained through study, however. No amount of pious exercise could create a muscle strong enough to grasp God's word. This wisdom would come only as a result of acknowledging one's sin and ignorance. It could come only through conversion. Fallen man had no way of coming to know God unless he allowed his hardened heart to be broken by the misery that comes with understanding his own corrupt nature and then regenerated by his maker into an instrument that would allow true communion. As God might grace the athlete with speed or the musician with perfect pitch, so he gives the contrite believer a heart that can understand the healing message of scripture. True communion did not happen at the cathedral's altar; rather, it occurred in the private space that came to be known as the heart.

This Protestant view of the heart finds its most memorable artistic expression in John Donne's Holy Sonnet XIV ("Batter my heart"), which was written around 1609. Donne, an Anglican priest, presents an image of a heart that is hardened against the sort of wisdom that will allow its owner to know and love God. Realizing that he cannot transform his heart on his own, the poet begs the Lord to "batter" it, to "break, blow, [and] burn" it so it can be rejuvenated. He also likens the heart to a besieged city.

> I, like an usurp'd town, to another due,
>> Labour to admit you, but O, to no end.
>> Reason, your viceroy in me, me should defend,
>> But is captived, and proves weak or untrue.

The heart needs to be overrun; it needs a new administration. Donne also compares the heart to being married to a wicked spouse and in need of a divorce. The poem ends with the apparently contradictory assertion that this new heart will be free only when it is imprisoned by God; that it will be chaste only once it has been ravished. For Donne, the heart is in critical condition. And it is only by the most radical intervention that the patient can be saved.

Donne's poetical formulation provides a fitting summation of an era that battered and broke traditional images of the heart, only to allow new, more durable ways of thinking about it to develop. Scientists, emboldened by new tools and methods, were finally able to strip away the fog of conjecture and superstition that for so long had prevented them from explaining the organ's anatomy and physiology. Artists and theologians, meanwhile, were able to stake out new metaphorical territory that would be able to resist the coming empirical assaults on the heart's emotional and spiritual sublimity. It is a dichotomy that lasts to this day when a man who has just undergone a heart transplant can, with a perfectly straight face, write a note thanking his doctors *from the bottom of his heart*—and everyone will know exactly what he means.

CHAPTER 4

Morbid Heart

Viareggio, Italy, 1822

The bodies have been in the water for more than a week. They finally wash up near Viareggio on the Tuscan coast. They are almost a mile apart. Although he is exhausted from seven days of frantic searching, Edward John Trelawny races up from Pisa the moment he hears the news, almost killing his horse from exhaustion in the process. Even so, he is too late. His dear friends have already been buried on the beaches where they were discovered. Quarantine laws are in effect, and nobody is taking any chances. The graves are marked with black handkerchiefs that are tied to driftwood sticks, as is the Italian way. Trelawny is forced to identify them by viewing a few waterlogged personal possessions found on the bodies of the two men who had been lost at sea. But these are proof enough.

As he travels back to Pisa to confirm the sad news to their gilded circle of English expatriates, Trelawny mulls over what to do. He knows he cannot leave the bodies where they currently rest. These are his friends. They deserve proper burials. *Splendid* burials, in keeping with their status as English gentlemen. And so, while everyone else mourns and cries and gnashes their teeth,

Trelawny goes into action. That is his role, after all. Naval officer, pirate—he is the man who does things.

But dealing with the Italian officials proves difficult. The threat of cholera lurks everywhere, and people are terrified of exposure to the dead. Trelawny remains undaunted. He bulls his way through all resistance, just as he had while seeking adventure on the Indian Ocean, and just as he would years later when he fought with Byron in Greece or swam the Niagara's rapids. In the end, he wins. It takes him more than three weeks of wrangling and threatening and cajoling, but Trelawny finally has his authorization to unearth and properly honor the lives of his friends.

In the meantime, he secures the equipment needed to build pyres on the beach, since transporting the bodies intact would be impossible. He constructs an iron frame, five feet long and two feet wide, that will hold the remains over the fire. And he has a coffin maker construct two small boxes covered in black velvet to hold the ashes. Each bears a brass plate with a Latin inscription giving the name and age of the man. The first:

Edward Ellerker Williams. Bengal army officer. XXIX anni.

And the second, the one that truly breaks Trelawny's heart:

Percy Bysshe Shelley. Poet. XXX anni.

Once the preparations are complete, he loads this sad cargo onto Lord Byron's yacht, the *Bolivar*, and sails up the coast. The great poet himself, claiming to be overwhelmed by grief, will come later by coach with another of their circle, Leigh Hunt. It is not an easy journey for Trelawny. This is the same boat he was on when the terrible event happened a month earlier. They had all traveled to Leghorn to meet with Hunt, who had just arrived in Italy from London. But Shelley had been eager to get back to Pisa to be with his young wife, Mary, who had recently miscarried. So Shelley and Williams had sailed back in the *Don Juan*, the small boat that Trelawny ordered built for them the previous year. Trelawny, the only able seaman among them, had planned to take the lead in Byron's larger vessel, the *Bolivar*. He was, as always, worried about his beloved friend Shelley. The man might be a genius, but he was also a hopeless captain—it was not uncommon for him to read Plato or Dante while at the helm. He also refused to hire local sailors for the journey, even though Trelawny would never dream of going out without them. What was worse was that he was without fear, which is never a good quality in a sailor.

The trip went wrong right from the start. The harbormaster had delayed Trelawny's departure with a prolonged inspection for contraband. By the time he made it to sea, the *Don Juan* had sailed away, subsumed by a skein of black clouds that hung like dirty rags over the smoky water. After ordering his mate to hasten after the other boat, there was nothing left to do but go below. To Trelawny's profound regret, he fell asleep. He was awakened by thunder; a sudden squall had materialized. He hurried on deck to help the boat's crew fight the swelling sea and whipping wind. All around them, other vessels labored toward shore. None of them was the *Don Juan*. The storm passed in a matter of minutes, as violent and transitory as one of Byron's fits of temper. Once it was gone, Trelawny desperately scanned the horizon, but this only confirmed what he already feared: Shelley's boat had vanished.

For the next week, he searched day and night for signs of the wreckage. He alerted coast guards from Genoa to Piombino. He pulled soldiers out of taverns and set them to work combing the beaches. He piloted the *Bolivar* into every narrow inlet and desolate headland. But all he uncovered were rumors. Some people claimed the *Don Juan* had been blown off course and found safe harbor in Corsica. Others said that the boat had been rammed by pirates looking for Byron's gold. A few even whispered that Shelley had used the opportunity to abscond with an Italian woman.

Most, however, believed that the *Don Juan* had capsized in the squall. Now, a month after Trelawny last saw his friends alive, the questions have all been answered, and he is on his way to give them proper funerals. Williams will be the first—his grave is the most accessible. His body is in a shocking state of decay; Byron, who travels up by coach with Hunt, can scarcely bring himself to look at it, though he does seize the opportunity to make a memorable remark, saying that Williams looks to be in worse condition than the black rag that was used to mark his grave. Then he swims out into the sea to challenge the waters that had stolen his friends. It is left to Trelawny to consign the body to flames. The searing heat does not take long to reduce it to ashes.

They cremate Shelley the following day. The spot where his body washed ashore is particularly desolate. The coast guard towers that stretch to the north and south are the only evidence of humanity. The black handkerchief has been blown away by the sea breeze, so it takes Trelawny several hours to locate the grave in the blistering-hot sand. As if possessing a sixth sense that allows him to avoid life's more squalid deeds, Byron arrives by coach moments after the body is found. He is accompanied by

Hunt, two mounted dragoons, and four foot soldiers. While the soldiers arm themselves with pickaxes, spades, and boat hooks, the dragoons use their horses to keep at bay the people who have arrived on several longboats. Some of them are well-dressed women. Word is out. An English poet is about to be cremated. They want to watch the spectacle.

Trelawny orders that spare sails from his boat be set up as breaks so the wind will not blow away Shelley's ashes. He then uses driftwood to build a pyre. Once the preparations are complete, he has the soldiers set about the grim task of disinterring the poet. They quickly reach the layer of lime that had been thrown on the body at the time of burial. And then a spade strikes something hard. A skull. Everyone stops, overwhelmed by the magnitude of what they are doing. The sound is loud enough for the onlookers to hear it; a woman gasps audibly. Trelawny urges the men on. The body is soon uncovered. As with Williams's, it is in a terrible state. The skin has turned black and hangs loosely from bones. Both of the legs are separated at the knees; the hands have fallen off. The skull is naked. There is nothing here of the man's beautiful countenance, over which, in his own words, the awful shadow of some unseen power had often passed. For a moment Trelawny allows himself to think that this is not Shelley after all, that some great mistake has been made and his friend is still alive. But then he sees something in the breast pocket of the corpse's jacket. A leather booklet. Two poems by Keats. Yes, this is Shelley. He was probably reading them when the storm struck.

Once the body is fully exposed, Byron steps forward and requests the skull. Trelawny attempts to pick it up for him, even though he secretly worries that the unruly lord will use it as a drinking cup—he has seen Byron do this before. But the skull immediately begins to break apart in Trelawny's hands and he is forced to leave it where it rests. After that, the soldiers take up the long, hooked poles that are used to drag bodies from the sea. Working carefully, they lift Shelley's brittle, rancid remains onto the iron frame. Trelawny then has them place two poles under it and, grabbing one end himself, they lay it all on the driftwood bier.

The fire catches quickly. The wood has been completely dried by the merciless sun. Trelawny and Byron throw frankincense, salt, sugar, and wine upon the body. Soon the blaze is immense, fueled by the ovenlike heat already rising from the sand. As the flames leap as high as the guard towers, Trelawny chants an incantation he wrote that morning especially for his lost friend, consigning his body back to the nature he had always worshipped.

Despite this inferno, the body burns slowly. Byron, acting bored to mask his horror, flees to the water to refresh himself, swimming out so far that some of the well-dressed ladies begin to call out in alarm. Trelawny keeps watch over the corpse as the outer layers of flesh turn to ash and crumble. Eventually the ribs crack open, revealing the heart. Even as the rest of the internal organs melt away, the heart remains stubbornly intact. The flames seem to dance around it without affecting its naked tissue. This is odd—no such thing had happened with Williams. Trelawny adds more fuel to the fire, but still the heart will not burn. A thick, oily substance sometimes leaks from it, creating a nimbus of white flame. But this only seems to shield the organ from the surrounding holocaust. Trelawny turns to the Italian soldiers—they, too, are astonished by the sight. So it is true. A great poet's heart is indeed extraordinary. The rest of his body may succumb to water and flame and time, but his heart, that cauldron of his immortal poetry, cannot be destroyed.

Knowing what he must now do, Trelawny orders the soldiers to carry the iron frame down to the sea. On his command, they partially lower it into the water, being careful not to soak the ashes. Great waves of hissing steam wash over what is left of the poet. The onlookers applaud and whistle, as if this is the climax of some tragic opera—which, Trelawny supposes, it is. He scoops up saltwater and sprinkles it on the heart to cool it, then takes the sizzling organ in his hand. It is still hot enough to burn him. He will not relinquish it, however. He wraps it in his handkerchief, then commands one of the soldiers to use the intact jawbone to scrape the body's ash into the specially made box. As the crowd of onlookers disperses, Trelawny takes possession of the box, keeping the wrapped heart and the brittle skull separate. As he leaves the beach, he notices a solitary bird flying back and forth over the dying fire, as if preparing to take Shelley's soul to heaven.

On the sad journey back to Pisa, he examines the heart more closely, thinking about what wonders it once contained. Or, perhaps, still contains. Lines of Shelley's poetry play through his mind, as do memories of the man himself, his brilliant conversation and ardor, his generosity as a friend. And yet, despite these lofty images, a more troubling thought starts to torment Trelawny as he looks at the charred organ. No matter how hard he tries, he cannot ignore the realization that this thing he holds in his hand is not transcendent at all. It is small and negligible. Viscid and shriveled. Charred and oozing.

It is ugly. Ugly and dead.

By the time he arrives in Pisa, he has managed to banish these troubling thoughts from his mind. He goes immediately to see Mary. This is his mission now, to give Shelley's immortal beloved her husband's miraculously intact heart. He is greeted by her friend Andrea Vaccà Berlinghieri, a professor of anatomy at the local university, who explains that the widow is resting in her bedroom. As they wait for her to be summoned, Trelawny shows the doctor the poet's heart and describes how it survived the intense heat of the pyre. Berlinghieri is less impressed than Trelawny had hoped. He responds, almost casually, that the heart is a particularly durable muscle. And in cases of suffocation, it becomes engorged with blood before death, which would have allowed it to resist the conflagration. Trelawny dismisses his remarks as characteristic of a man of science, a fact lover who has never experienced sublimity. The doctor seems more interested in the skull, remarking that he has never seen one so extraordinarily thin. Mary Shelley overhears this last remark as she enters. In deference to the widow, Berlinghieri adds that this is no doubt due to the unique sensitivity of the brain it contained.

After describing the extraordinary events at the beach ceremony, Trelawny then presents Mrs. Shelley with the poet's heart. Her reaction is not what he expects. The author of *Frankenstein* looks at it with terror and disgust. She gestures for him to remove it from her sight, looking for a moment as if she might even swoon. Then, in a weak voice, she asks that it immediately be taken with her husband's remains to Rome to be buried in the Protestant Cemetery, according to his wishes. Trelawny, remembering the recent miscarriage that had sent Shelley to sea and his doom, understands that giving her the heart might not have been the flamboyantly gracious gesture that he had intended. He readily agrees to her request, though he will disobey her in one regard: he will keep possession of the heart. He does not have it buried. Later, when the storm of emotions has passed, he sends it to her. This time she does not reject it. And in fact, she keeps it in her desk until the day of her death, though it is doubtful that she ever takes it from its box.

The members of the exalted Pisan circle quickly go their various ways, leaving the interment of Shelley's ashes in the hands of the British authorities in Rome. Trelawny spends the autumn and winter shooting on the Italian coast, Mary returns to England, and Byron . . . well, he continues being Byron. Early the following year, Trelawny travels to Rome to visit the grave of his beloved friend. There, he is disgusted to discover that Shelley's remains are placed among those of commoners and vagabonds. He soon arranges for the remains to be moved to a more fitting spot, a verdant elevation near the great pyramidal tomb of some ancient Roman figure. In addition to a quotation from Shakespeare, he has a Latin phrase inscribed on the gravestone: *Cor cordium*.

"Heart of hearts."

THE heart got sick in the eighteenth and nineteenth centuries. Now that the basic elements of its form and function had been established, the myocardial muscle was no longer an enigma. It could run, but it could no longer hide. There was, of course, much still to be discovered about the particulars of its anatomy and physiology. There would be no more decisive breakthroughs, however— nothing that would shake the foundations of humankind's understanding the way Harvey's work had. The work of the French anatomy professor Raymond de Vieussens, who in 1706 provided the first detailed map of the heart's chambers and major vessels, meant that it was now possible to hold an accurate image of the heart in the mind. And, as with the drowned poet's heart, it was not always a pretty sight. It could be damaged. It could be diseased. It could be *deadly*. Now that physiologists understood how the heart worked, they wanted to understand why it often did not—and what, if anything, they could do about it.

This proved a difficult transition. One of the most enduring fallacies of the medical profession, dating all the way back to Hippocrates, was that the heart was immune to disease. Right up until the end of the eighteenth century, many physicians still believed that heart ailments were relatively rare, and perhaps even nonexistent. Symptoms we now associate with coronary disease—such as fainting spells, breathlessness, and crushing chest discomfort—were believed to have nothing to do with the heart. Even sudden cardiac death was viewed as a normal cessation of the muscle's activity, a giving out of its allotted reserve of vitality. The notion of heart *disease*—the central health concern of our time—simply did not exist.

It was during the eighteenth century that a pioneering group of doctors began to challenge the prevailing wisdom. Like rogue detectives and crusading journalists, they robustly interrogated the heart, asking questions of it that their predecessors would have thought absurd. Their interactions with the heart were no longer loftily reverential, but rather grittily diagnostic. They saw sick people and suspected that the heart was the culprit. Demystified, dethroned, and defrocked, the heart could now be probed, palpated, and tested. It was expected to make an accounting of itself. By the beginning of the nineteenth century, its responses were providing researchers with enough data that they could start naming the heart's diseases. Terms like *pericarditis*, *angina*, *arteriosclerosis*, and *endocarditis* entered the medical lexicon. The heart had become morbid.

To conduct their investigations, researchers exploited the rapidly expanding technologies of the Industrial Revolution. For the first time, instruments and machinery were brought to bear on the beating heart. Early attempts were inevitably crude, such as when, in 1665, the British anatomist Richard Lower used a simple goose-feather quill to attach the arteries of two dogs and performed the first known blood transfusion. Two years later, Lower used the same process to transfuse a sheep's blood into a mentally ill man in an attempt to cure him. Needless to say, the operation was not a success. Eventually the gadgetry became more sophisticated. As the incursion deepened, pathologists began to look at the heart not so much as a paragon of divine design, but rather as a flawed mechanism liable to break down in an alarmingly broad number of ways.

By the middle of the nineteenth century, everything had changed. The heart was now widely seen as a source of death as well as life. Today's "number one killer" had been born, as had a new discipline, cardiology, and a new breed of physicians who focused primarily on the heart. Many of these newfangled specialists worked in hospitals dedicated to the treatment of cardiac disease, such as London's National Heart Hospital, which opened in 1857 and was still thriving 111 years later, when it was the site of England's first heart transplant.

With the heart coming to be viewed as an engine of disease as well as a fountain of life, public worry about it grew exponentially until, by the middle of the nineteenth century, heart anxiety was epidemic. This disquiet was reflected in the era's literature and poetry. While previous generations of writers had limited their representations of the heart's destructive influence to the emotional upheavals it could cause and the evil impulses it might harbor, novelists and poets now had at their disposal a whole new way of thinking about how the heart might attack the individual. It became a physical threat. It could strike suddenly, without warning. Or it could slowly sap the life from a person. The Victorian era's assembled literary characters came to resemble patients in a vast sick ward, with the diseased heart vying with tuberculosis and nervous disorders as the malady of choice.

This new era needed a new discipline. "Morbid anatomy" was created by the mid-eighteenth-century Italian physician Giovanni Battista Morgagni, who is generally acknowledged to be the founder of modern pathology. As its name suggests, this branch of study views the human body as a crime scene rather than a paragon of animals. Although Morgagni did not focus on the heart specifically, his methodology set the stage for a new way of viewing the cardiac muscle. He was the first to examine a cadaver with an eye toward tying what he uncovered with the symptoms of disease recorded during the person's lifetime. The stilled heart was no longer seen just as something that could shed light on how its owner had lived. Rather, it could potentially explain how he had died.

As Morgagni was advancing the practice of morbid anatomy, an Englishman named William Heberden was identifying a condition that would play an increasingly important role in the diagnosis of heart disease. In 1768 Heberden delivered a paper at London's Royal College of Physicians that described patients who suffered crushing sensations in their chests in response to physical effort. In almost every case, this pain was alleviated by rest. He labeled this condition angina pectoris, which conflates the Greek word for strangling, *ankhone*, with the Latin *pectus*, or chest. (This is the syndrome that afflicted our imaginary Greek merchant, Nikias.)

Although Heberden did not initially associate this pain with heart disease, his colleagues soon began to connect the dots. Symptoms such as sharp, radiating pain along the left arm; episodes of fainting; and a generalized feeling of impending doom were culled from a wide array of cases in several nations and added to the description of the syndrome. Breathlessness also came to be part of angina's symptomatic suite. Eventually, using Morgagni's technique of tirelessly registering clinical cases and autopsy results, doctors determined that sufferers of angina tended to have something wrong with their hearts. The question—which would not be answered for two centuries—was precisely *what* was wrong. Some believed (correctly, it turned out) that it was what they called ossification of the vessels surrounding the heart. Others suspected dropsy (swelling) in the pericardium, the leathery sac that encases the heart. Still others considered the condition to be caused by abnormalities of the heart muscle itself, such as hypertrophy (enlargement). Whatever the cause, an important shift had begun. Symptoms long seen in a broad population of patients were now attributed to the heart.

A disorder that would eventually be seen as one of the main

causes of angina was also beginning to be better understood. In 1833 a surgeon named Johann Lobstein began using the term *arteriosclerosis* to describe a condition in which fatty deposits gathered on the inner walls of arteries, constricting and hardening them. Although some doctors were already making speculative connections between this condition and a fat-rich diet, there was, as of yet, no treatment for the disease. Nor would there be for more than a century, when bypass surgery, balloon angioplasty, low-fat diets, and statin drugs were deployed to lessen the chances of catastrophic blockage.

Congenital heart defects were also beginning to be more widely recognized. There was no longer any denying that some hearts were born different from others. Harvey's perfectly proportioned king could instead, like Shakespeare's Richard III, possess a congenitally crooked back and nasty disposition. True, since ancient times there had been recorded instances of a condition known as ectopia cordis, in which a baby is born with its heart on the outside of its ribs. But these were very rare and usually attributed to supernatural causes. With the increasing sophistication of human anatomical investigation, however, a variety of congenital heart defects—shunts and holes and malformations—was identified, the sum total of which suggested that the heart was not the Platonic ideal of perfection once believed. Once again, however, these discoveries led only to the frustrating realization that there was precious little a concerned doctor could do to cure them.

The news was not all bad, however. At least the heart could be made to speak to the attentive physician. Trying to make sense of the heart's sounds had been an important part of medical inquiry ever since the time of the ancient Egyptians. Galen was particularly adept at feeling the pulse, extensively cataloging at least twenty-seven types of heartbeats. Without the possibility of postmortem examination, however, these sounds had always been little more than confounding chatter.

The first great breakthrough in deciphering the heart's babel came in the 1760s at the hands—literally—of the son of an Austrian innkeeper. Remembering how his father would rap on large wine casks to determine their volume, Josef Leopold Auenbrugger came up with the idea of tapping on the chests of his patients to assess the fluid levels of the organs and vessels inside. Percussion, as it was called, gradually became one of the foremost weapons in the diagnostic arsenal. In order to test this method, Auenbrugger assiduously conducted a large number of pathological examinations on the dead bodies of patients he had percussed while they were alive, contrasting the actual fluid levels with those he had reckoned by ear. The result was a remarkable catalog of heart sounds. His most radical and ingenious inquiry involved filling the pleural cavity—the membranous sac containing the lungs—of a cadaver with fluid before he sounded it. Although Auenbrugger published his results, his methods were ignored for almost half a century, until they were taken up and popularized by the great French professor Jean-Nicolas Corvisart.

Percussion, however, had its limits, not the least of which was the skill of the examining doctor. In fact, during this time, training in the discipline that would soon come to be called cardiology often resembled a musician's tutelage, with students being graded on their manual dexterity and aural acuity. An even greater leap forward in decoding the heart's sounds came in the early part of the nineteenth century, when Parisian physician René Théophile Hyacinthe Laënnec had a brainstorm after watching children playing with a log in the courtyard of the Louvre. When one group of children scratched one end of the log, the others, pressing their ears to the wood at the far end, were able to hear what the naked ear could never have detected. Laënnec, a flutist who had been trained to listen to the heart by both percussion and direct auscultation (placing his ear on the chest itself), had long been frustrated by the limits of these methods, particularly with the obese. Female patients, meanwhile, demurred at having a young male doctor touch their chests (at least publicly). As Laënnec writes in his seminal work, *Mediate Auscultation*:

> In 1816, I was consulted by a young woman laboring under general symptoms of diseased heart, and in whose case percussion and the application of the hand were of little avail on account of the great degree of fatness. [Direct auscultation] being rendered inadmissible by the age and sex of the patient, I happened to recollect a simple and well-known fact in acoustics . . . the great distinctness with which we hear the scratch of a pin at one end of a piece of wood on applying our ear to the other. Immediately, on this suggestion, I rolled a quire of paper into a kind of cylinder and applied one end of it to the region of the heart and the other to my ear, and was not a little surprised and pleased to find that I could thereby perceive the action of the heart in a manner much more clear and distinct than I had ever been able to do by the immediate application of my ear.

And so the stethoscope was born. Laënnec soon built the first dedicated device, a hollow wood tube a foot long and about an inch in diameter. It caught on as quickly as a hot new software application would in our time, undergoing a series of refinements until, by century's end, the current binaural device—familiar to anyone who has ever had a physical exam or watched television—was in place. Besides having obvious clinical applications, the stethoscope also became a favorite toy in parlor games, with men and women achieving hitherto forbidden intimacy under the guise of scientific education.

Armed with his new tool, Laënnec set about giving names to the murmurs he could now clearly hear. He proved himself to be worthy of his poetic name. For the sound that occurs when the atria contract, he proposed the term *bruit de rape*, which translates as a file grating on a piece of wood. If the murmur sounded like a soft breath, as in stenosis (narrowing) of the mitral valve, he called it a *bruit de souffle*. The most beguiling term that Laënnec coined was his description of the tremor that could accompany a murmur associated with ossification or stenosis of the aortic valve: *frémissement*, a "purring or shivering" of the heart. As expertise improved, the heart's sounds became an ever-more-important diagnostic tool. The whispers that echoed through its chambers were every bit as important—and potentially suggested scenarios as lethal—as gossip at a royal court.

THIS new technology found a powerful echo in the work of an author who was very much in tune with the era's increasing interest in morbidity—Edgar Allan Poe. Throughout his writing, Poe expressed a deep unease with science's desire to probe the human organism for disease and defect. His 1829 poem "To Science" expresses distrust of the "peering eyes" of science, which, he claims, deal only with "dull realities." He goes so far as to compare science to a vulture that feeds upon "the poet's heart."

Poe was very much aware that the heart was no longer seen as solely the wellspring of love and life. It was also an agent of terrors that could be uncanny, even unspeakable. In his short story "The Premature Burial," the heart plays a vital role in bringing about the fulfillment of one of humankind's deepest nightmares. The story's cataleptic narrator describes several cases in which the apparent termination of a pulse causes a person to be buried alive:

> We know that there are diseases in which occur total cessations of all the apparent functions of vitality, and yet in which these cessations are merely suspensions, properly so called. They are only temporary pauses in the incomprehensible mechanism. A certain period elapses, and some unseen mysterious principle again sets in motion the magic pinions and the wizard wheels.

In one such case, disaster is forestalled only by the use of a battery applied directly to the pectoral muscle. In Poe's work, the heart becomes a trickster capable of leading us to unspeakable horrors.

His use of the heart as an instrument of dread finds its deepest expression in his masterful 1843 short story "The Tell-Tale Heart." The story opens with an unnamed narrator confessing to the murder of an old man with whom he has been living. In his defense, the killer claims to have been suffering from a "disease" that heightens his senses, particularly his hearing. It also makes him acutely sensitive to the old man's clouded blue eye, which disgusts him so deeply that he calls it "the eye of a vulture."

After stalking his victim for a week, the narrator finally murders him. He then dismembers the old man—using a tub so there will be no bloodstains—and hides the remains under the floorboards. He appears to have committed the perfect crime. But then the police arrive, alerted by a neighbor's report of a terrified shout. Confident that he will never be caught, the killer blithely shows the police around the house, eventually leading them directly to the scene of the crime. His poise is soon shattered, however, when he begins to hear the old man's heart beating beneath their feet. The police display no sign that they detect anything amiss, though the narrator thinks this is simply a ruse. He soon cracks and confesses.

The story is a masterpiece of gothic horror, one of the most profound explorations of madness ever written. To achieve this effect, Poe cunningly taps into his readers' growing unease about their own hearts, whose workings can now be laid bare to any doctor with a stethoscope. On the night of the murder, as the narrator observes his prey, his hypersensitive ears become aware of "a low, dull, quick sound, such as a watch makes when enveloped in cotton. I knew *that* sound well, too. It was the beating of the old man's heart." This "hellish tattoo" infuriates him as it grows louder and quicker. "I thought the heart must burst," he claims just before he pounces. As he smothers the old man, "the heart beat on with a muffled sound." It is only when this dreadful noise ceases that the narrator can be sure that he has succeeded.

And yet "the beating of his hideous heart" soon begins again, leading the killer to his doom.

Critics and readers have long been divided over the intended source of this heartbeat. Those of a supernatural inclination think that the old man's heart lives on outside his mutilated corpse. Others attribute it to the rhythmic click of the death-watch beetles that likely infest the room's walls. Most, however, think that the conscience-stricken killer is actually hearing the report of his own heart as it echoes in the canals of his oversensitive ears. This last interpretation makes the most sense. Laënnec's recently invented stethoscope would have been in wide use by the time of the story's composition. It would have been common knowledge that merely listening to a heart could divulge its inner secrets. The heart could tell tales about its owner, tales of disease and impending death.

It is worth noting that the old man's evil eye is compared to a vulture's—the same image used to characterize science in Poe's earlier poem. The old man represents the spirit of scientific inquiry that threatens to rob the heart of its sublimity, while the narrator is the beleaguered poet, in danger of being picked clean by the physician's tools.

Laënnec's rolled sheets of paper may have been rudimentary in the extreme, but this was, in fact, not the first instance of a physician using technology to probe the heart. This honor must go to a contraption invented in the early seventeenth century by Santorio Santorio, also know as Sanctorius of Padua, an Italian physiologist whose name suggests that he was as much magician as doctor. It is easy in hindsight to view Sanctorius as something of a crank. For instance, in an effort to determine how much food the body burns, he used an ingenious "weighing chair" to measure everything he ate and drank against all the waste he produced—for a period of thirty years. Needless to say, the outgoing sum was consistently smaller than the incoming, leading Sanctorius to theorize that the difference evaporated in a sort of undetectable perspiration. While his hypothesis might be easy to dismiss, his methodology displays the single-mindedness without which scientific breakthroughs rarely occur.

Sanctorius's greatest creation was a device for measuring the heart rate—the pulsilogium. This consisted of a pendulum attached to the end of a ruler. When the pendulum's swing was synchronized with the patient's heartbeat, the operator was able to measure its arc against the ruler, thereby determining the frequency of pulsation. Of course, this raises the question of why the operator did not simply *count* the number of beats, but the device's crudity should not blind us to its groundbreaking theoretical importance: machinery could now measure the heart.

An equally industrious cardiotechnician was the English inventor Stephen Hales, who attempted to take the first mathematically accurate measurements of the heart's pumping capacity. His 1733 book, whose name, *Haemastaticks*, evokes a world in which the principles of mechanical engineering are being applied to the human body, details a series of ingenious if stomach-turning experiments to measure blood pressure, as well as the heart's rate of flow and its capacity. To do this, he attached a catheter made from a goose's trachea to an eleven-foot-long glass tube. He then inserted the tube into the carotid artery of a still-living horse. As the animal bled to death, Hales tracked how far up the tube the horse's heart was able to project the blood with each pulsation, using this as the basis of the first measurements of arterial pressure. Hales also inserted a gun barrel through the punctured neck and into the heart of a recently slaughtered sheep and then poured molten wax through the barrel to make casts of the heart's chambers, which allowed him to determine their volumes. Although his tabulation of these results suffered somewhat from his inability to perform similar experiments on human beings (he was, after all, an ordained curate), Hales began a tradition in which the heart was subject to the same quantification as a hot-air balloon or steam engine. Harvey's bodily king had been transformed into a yeoman, whom physiologists had every intention of putting to work.

Blood pressure gauges that could be used without causing a slow, agonizing death were developed in the next century. In the 1840s Carl Ludwig invented the kymograph. This instrument consisted of a rolling drum that was covered with a sheet of paper and used a stylus that was directly attached to an artery to record fluctuations in pressure. In the following decade, another German, Karl von Vierordt, invented the sphygmograph. This cumbersome device looked nothing like the current sphygmomanometer, with its cuff, inflator bulb, and gauge. Rather, it consisted of a suitcase-sized base plate on which was mounted a triangular wood frame with a perpendicular apical bar, from which two horizontal rods were suspended. These rods, which supported a number of pans, were fastened to a pulley, whose other end was wrapped around the arm in a manner similar to the contemporary cuff (which anyone who has had his or her blood pressure taken by a doctor or nurse will know). Weights were then gradually added to the pans until blood flow to the arm's artery was completely cut off, thereby providing a measure of the blood pressure.

The device was improved upon ten years later by the French physician and photographer Étienne-Jules Marey, whose famous stop-action photo series of men jumping, horses galloping, and birds taking flight played an instrumental role in the development of cinematography. Marey's contribution to cardiology was a portable sphygmograph that could be attached to the wrist. It resembled an arm guard used by archers. The main mechanism was a metal lever, one end of which was set against the artery on the patient's wrist. The other end held a stylus, which would trace the blood vessel's varying pressure on a scroll of paper that a clockwork mechanism threaded evenly through the machine. The contemporary sphygmomanometer, which uses a mercury meter to record blood pressure, was developed two decades later by Austrian Samuel von Basch and subsequently improved by the Italian pediatrician Scipione Riva-Rocci.

THE most important technological advances of the era involved the effects of electricity on the heart. Scientists had known that a body could be in some way electrified since the ancients had observed the often deadly stings given by eels and other types of marine stun guns. An ancient Roman physician named Scribonius Largus even suggested that the discharge of an electric torpedo fish might have some benefit in relieving chronic pain. Seventeen hundred years later, in 1785, Charles Kite, a founding member of the Society for the Recovery of Persons Apparently Drowned, prefigured the modern defibrillator when he proposed the use of two brass wires attached to a generator to revive drowning victims.

The electrified heart came into its own in the 1780s, when the Italian physicist Luigi Galvani touched the sciatic nerve of a frog he was dissecting with a scalpel charged with static electricity. There was a spark, and the animal's leg jumped. After carrying out a number of ingenious experiments, Galvani concluded that nerves use electrical current to move the body's muscles, a process he called animal electricity. From here, it was an easy step to conclude that the application of electricity to the human body could stimulate its most powerful and active muscle—the heart. This notion was bolstered a few years later by Galvani's nephew, Jean Aldini, whose experiments on criminals in London, carried out immediately after their executions, established that muscles could be set in motion by electrical charges for up to two hours after death had been declared. Galvanism, as the process came to be known, would come to play a role in some of the nineteenth century's bolder, more outlandish cardiac interventions.

At about the time Aldini was pumping electricity into the corpses of hanged men, a woman was born in London who would make unforgettable use of these new theories of animal electricity. When the woman who would become Mary Shelley was only eighteen, during a vacation at the Villa Diodati on Lake Geneva, she listened closely as her lover, Percy Bysshe Shelley, and his friend Lord Byron discussed Galvani's and Aldini's research. As it happened, Percy Shelley, while a student at Eton, had actually seen a teacher re-create Galvani's frog experiment. These discussions were very much on Mary's mind later during the rain-blighted holiday when Byron proposed that they pass the time by each conjuring up a ghost story. Although the men abandoned the contest to go hiking once the lightning storms had cleared,

Mary, who was recovering from a miscarriage, stayed behind to write. The result was a short novel that captures the era's growing anxiety about science and morbidity every bit as unforgettably as Poe's work would a few decades later.

Frankenstein, published in 1818, is deeply informed by the ongoing revolution in medical technology. The woman who would later turn away in disgust from her husband's charred organ is clearly alarmed by the breakneck advance of science. Shelley's book is not just a cautionary tale of scientific hubris, however. It also presents an evocative portrait of the heart's endurance as a metaphor even as technology demystifies the body.

Shelley's hero, Victor Frankenstein, would have considered himself very much a peer of Hales, Galvani, and Aldini. After an early infatuation with alchemy, the young Swiss student places himself squarely in the scientific vanguard through an almost superhuman act of ambition and will. The monstrous "daemon" he creates in his garret room is also a logical extension of the work of the Renaissance anatomists. What they took apart, Frankenstein stitches together again. Although he initially grumbles about having "to dabble in dirt" by peering into microscopes or laboring over crucibles, he soon realizes that chemistry and biology—or natural philosophy, as he calls them—will allow him to perform the sort of life-engendering miracles he has always sought.

Frankenstein never specifies exactly how he jump-starts his creature's heart. Most of Shelley's contemporary readers, well aware of Galvani's work, would have reached the same conclusion—that it involved electricity. What Frankenstein does confess is that making monsters is a gory business. "It was, indeed, a filthy process in which I was engaged," he claims. "I went to it in cold blood, and my heart often sickened at the work of my hands." In fact, when the adventurer Robert Walton, who hears Frankenstein's tale when he finds him wandering in the icy wilderness, asks him about "particulars of his creature's formation," Frankenstein pointedly refuses to divulge them, in the hope that they will be forever forgotten. It is left for the reader to guess whether the heart was transplanted whole from an unwitting donor or assembled from parts culled from the "dissecting room[s] and slaughter-house[s]" Frankenstein secretly raided.

Whatever its origin, the monster's heart is physically potent enough to provide him with great speed and endurance. It is also vast in its need for affection, its capacity for vengeance, and, ultimately, its accommodation of a remarkably powerful conscience. Despite its size, however, this heart is still all too human. When first loosed upon the world, the monster finds food, clothing, and shelter to be secondary requirements. What he really needs to quench is his heart's immense thirst for human love and contact. He confesses that, after secretly observing the De Lacey clan at a cottage he happens upon, "my heart yearned to be known and loved by these amiable creatures: to see their sweet looks turned towards me with affection, was the utmost limit of my ambition." He also craves a wife, less for physical release than for "the interchange of those sympathies necessary for my being."

When he is spurned by both the De Laceys and Frankenstein himself, the monster's pain and fury are intense. He grows to hate his "heartless creator" for giving him human sentiments but encasing them in a body that provokes horror and disgust. His heart warps, but along human lines. He soon becomes a killer. His crimes stem from a surfeit of human emotions, however, not a lack of them. His most heinous murder, of Frankenstein's six-year-old brother, William, is an affair of the heart. "I gazed on my victim, and my heart swelled with exultation and hellish triumph." His actions are not those of an unthinking beast or an affectless psychopath, but rather of the only creature on earth capable of vengeance: man.

The monster's heart is quantitatively, not qualitatively, different from those of the humans he meets. Here we see something of the same mistrust of science that Poe was to show a few decades later. For Shelley, scientific intervention makes the heart outsized, overly powerful, capable of killing. Like his eight-foot frame, which can climb mountains and run with the speed of a gazelle, the monster's engineered myocardium is terrifying because it is too strong, too feeling, *too human*. Couple this with the rejection that comes from possessing a hideous face and body and it is hardly surprising that a one-man crime spree ensues.

The rampage takes an emotional toll on the killer. After ultimately destroying Frankenstein, the monster is stricken by a powerful attack of something that looks very much like conscience. When Walton, who encounters him after Frankenstein's death, chastises him for not listening to that still voice in his heart before hunting down Frankenstein, the creature begs to differ. "My heart was fashioned to be susceptible of love and sympathy; and when wrenched by misery to vice and hatred, it did not endure the violence of the change, without torture such as you cannot even imagine." So it was not just the monster's capacity for love and vengeance that was outsized, but also his ability to feel remorse. One of the book's deepest ironies is that the reader is left to wonder whether Victor Frankenstein could have claimed such a powerful conscience.

Frankenstein has been rightly celebrated as a cautionary tale about scientific hubris, a book that peered over the horizon to our own era of genetic engineering, cloning, and transplant technology. When it comes to matters of the heart, however, Shelley was very much in tune with the Romantic poets with whom she was so intimately involved. The book's metaphorical use of the heart, even when the monster himself is speaking, remains strictly conventional. He describes his heart "sinking" when he is rejected; he characterizes it as "softening" when he envisions a rapprochement with his creator. In the end, when it comes to imagining the deformities brought upon humankind by the scientific revolution raging around her, Shelley's otherwise peerless imagination was only skin-deep.

GALVANI'S experiments inspired not only Shelley. They also proved the basis for the work of a dynamic doctor named Augustus Desiré Waller as he set about finding a means of measuring the electrical activity in the human heart. In the 1880s Waller created a machine he called the electrocardiogram, though it was in fact only a crude precursor to what we now know as the ECG—electrocardiogram—machine. (It is also referred to as an EKG.) Suspecting that the currents that cause the heartbeat could be recorded, he set up a mechanism in which a subject's limbs would be placed in basins of saltwater. The patient was then connected to an electrometer, which was rigged to project an image of the heartbeat onto a photographic plate that had been fixed to a passing model train. The resulting graph showed a wave pattern that he called an electrogram—a telegram from the heart. However, although he proved that the heart generated a pattern of electrical activity, his machinery was not sensitive enough to be of diagnostic value. The telegrams could not be decoded.

Waller popularized his findings in a series of lectures that famously began with the words, "I am going to describe how the heart of man can be shown to act as an electrical organ." In keeping with the cavalier spirit of the time, he used both himself and his long-suffering bulldog, Jimmy, to demonstrate this process. For many in his audiences, the notion that their cardiac muscles were powered by electrical currents—that their hearts were electric—must have come as something of a shock indeed.

Waller's pioneering efforts would bear ripe diagnostic fruit in just more than a decade, when the Dutch physiologist Willem Einthoven devised a way to accurately measure the cardiac muscle's electrical activity. In 1902 Einthoven invented a device called a string galvanometer, which consisted of a series of long, extremely thin, silver-coated glass filaments, or wires, capable of conducting electricity generated by the heart without interference or distortion. The creation of wires fine enough to meaningfully record the heart's electricity came only after Einthoven had the idea of attaching molten glass to an arrow, which he then had fired from a bow across his laboratory, stretching the glass into extremely fine filaments. He then situated these wires inside a powerful electromagnetic field. When the wires were attached to a patient, the magnets would bend them to different degrees according to the cardiac current they carried. This displacement was projected onto a photographic plate, creating a spiked graph. The ECG machine, perhaps the single most important diagnostic tool in the history of cardiology, was born. Although these machines are now relatively small, Einthoven's original device weighed more than five hundred pounds because of the size of the magnets. With it, the physician was no longer reliant solely upon his ear to try to figure out what the heart was saying to him. He now had access to a more profound language, one emanating from the muscle's every cell. The heart's deepest code was finally being cracked.

Investigations into the heart's electrical activity raised important questions: Who switched it on? What actually caused a heart to beat? Was the origin myogenic—in the muscle itself? Or was the heartbeat neurogenic, coming only after instructions were issued by the brain? From the early Greeks right through to Harvey, it had been held that the heart's beating was fundamentally myogenic, that it originated in the tissues and chambers of the heart itself. This view reached perhaps its most sophisticated articulation in the middle of the eighteenth century with the work of the Swiss anatomist, botanist, and poet Albrecht von Haller, who maintained that muscles were fundamentally "irritable"—that is, their tissue possessed an innate quality that caused them to react when aggravated by external stimuli. Haller saw the heart as the body's most irritable organ (a view to which generations of lovers, poets, and playwrights heartily subscribe). When it dilated, blood rushed into its chambers. According to Haller, this turbulence stimulated the heart's muscular fibers, causing them to flex and contract. This violent systolic action then propelled blood on its long, nourishing circuit through the body. When it returned, the process of stimulation and reaction repeated. The brain, he said, had nothing to do with it.

Almost immediately, anatomists began to undermine Haller's theory as they mapped the nerves running from the brain to the heart. They maintained that it was these, not the blood vessels, that stimulated that prickly myocardium. The neurogenic theory was born. The heart beat only upon the brain's command. But then, in the latter part of the nineteenth century, the English physiologist Walter Gaskell performed a number of experiments on the hearts of tortoises that seemed to demonstrate that the heartbeat's origin—its *activation*—came from within the muscle itself, though it could also be powerfully affected by impulses coming from the brain via a bundle of nerves at the top of the heart.

In the end, a sort of cease-fire was reached between the two camps. As we now know, Gaskell was correct. Both theories are accurate. The heart is richly innervated, having connections to both the sympathetic and parasympathetic nervous systems. Its rate is governed by the vagus nerve, also known as the tenth cranial nerve, the Wanderer, and the Rambler. This nerve runs from the medulla along the jugular vein to the heart, where it instructs the sinus node and the AV (atrioventricular) node to speed up or slow down the pulse. But the heart is also capable of generating a beat without the input of the nervous system, in a process known as the action potential. Each heart cell's membrane has an active pump that controls a complex exchange of sodium and potassium across it, resulting in a small electrical impulse. When added together, these impulses cause the heart to beat.

At the time that this debate was raging, an American surgeon was investigating a different sort of irritability, one that would provide a view of the heart that would have enormous ramifications in the following century. In an article published in 1871,

Jacob Da Costa discussed an ailment that he and other doctors were observing in soldiers, particularly those who had seen battle in the American Civil War. Its symptoms looked very much like angina: difficulty breathing after strenuous effort, palpitations, anxiety, and chest pain. The only difference was that the heart attacks that so often followed bouts of angina were not happening. The problem was known as both irritable heart syndrome and soldier's heart syndrome. British doctors who had witnessed the same phenomena in the Crimean War thought they had an explanation: the symptoms were caused by the straps of heavy knapsacks that constricted the heart when fastened across the chest.

Da Costa was not buying it. In his experience, soldiers who did not carry heavy weights were also presenting with these symptoms. Search as he might, however, he could find no evidence of underlying disease or physiological disorder. Finally, after closely studying three hundred Civil War veterans, he concluded that their pain and discomfort were caused by what we now call emotional stress, albeit of a severe variety. Treatment was rest and, when possible, removal from the heat of combat. In our era of battle fatigue and post-traumatic stress disorder, this might seem like little more than common sense, but Da Costa's understanding of the connection between stress and heart disease—or at least its symptoms—was revolutionary.

His conclusions could not have been more timely. Cardiac anxiety was now epidemic within the general population. Although the nineteenth century has often been called the Century of Nerves in reference to the number of people who suffered from nervous disorders, it might just as easily have been called the Time of Palpitations. Everybody seemed to be having heart trouble. Deaths that had formerly been attributed to respiratory ailments or communicable diseases were suddenly credited to heart failure. In hindsight we can see that many of these attributions were erroneous. But that did not lessen their hold on the greater imagination. Women—and men of a sensitive, poetical disposition—were seen as particularly prone to cardiac difficulties. As the cultural critic Kirstie Blair points out in her fine study on the topic, *Victorian Poetry and the Culture of the Heart*, it is almost impossible to find a Victorian poet who did not complain of heart ailments at some point in his or her poems, journals, or letters. Symptoms ranged from occasional palpitations to "cardiac apoplexy," the era's preferred term for a heart attack. When artists and lovers claimed to be heartsick, they were no longer simply employing another word for melancholy. They were referring to a specific ailment, a degeneration of cardiac function brought about by acute nerves and an inability to manage emotional turbulence. By midcentury, nearly every swoon, gasp, and chest ache was attributed to a sick heart by a public whose imagination was stoked by poets, novelists, newspapers—and a growing segment of a medical community that was all too ready to tell the sufferer what she or he wanted to hear.

It was not, of course, all doom and gloom. The metaphorical heart continued to contain positive qualities during this period. Indeed, two relatively new notions were incorporated into the heart's symbolic repertoire: conscience and sympathy. The idea of conscience, of course, had been around for as long as people had pondered their moral lives. In Christianity it was usually defined as an individual's ability to distinguish God's word from Satan's—to do what our maker wanted us to do. During the Age of Enlightenment, philosophers challenged the exclusively religious nature of conscience. Thinkers as different as Immanuel Kant and Jean-Jacques Rousseau claimed that conscience was an inherently human quality whose existence did not depend on a deity. This shift did not evict conscience from the heart, however. If anything, it strengthened the heart's role as our moral crucible. Rousseau, that most emotional of philosophers, saw the heart as both the source of our passions and the arbiter of our moral lives. The wise man listened to his heart in matters of both love and duty. The trick was to harmonize the two roles. "A feeling heart . . . was the foundation of all my misfortunes," Rousseau famously states early in his *Confessions*, indicating not only the difficulties brought about by his pursuit of the heart's passions as a young man, but also the trouble caused by adhering to the dictates of his conscience. Once again we see the heart's deep metaphorical durability. No matter how often it is evacuated, meaning keeps creeping back into those empty chambers.

The other new metaphorical use of the heart was as the engine that generated our fellow feeling, our *sympathy*. It was the organ of compassion. It was through the heart that a person could feel another's joy or, more acutely, his or her suffering. The heart was the link that fastened us to an infinite web of human feeling. In the famous tenth stanza of William Wordsworth's "Ode: Intimations of Immortality from Recollections of Early Childhood" (1802–1804), the poet reflects on a child's ability to feel the "gladness of the May" in his young heart. Although he acknowledges that "nothing can bring back the hour / Of splendour in the grass" now that he is an adult, he will not grieve, taking strength from the "primal sympathy / Which having been must ever be." Through its capacity for sympathy, for recollecting what children feel, the heart ties us into a happiness that cannot be diminished by absence or the passage of time. A feeling heart not only is the foundation of our conscience, it also gives us our capacity for brotherhood. It is our membership card in the family of man. As other writers of the era remind us, however, this finely tuned receptivity can also backfire on the individual to become a source of considerable derangement. The image of the oversensitive poet, dying in his garret from heart disease or tuberculosis because his raw nerves could not handle a cruel, buffeting world, was so common that it ultimately became a cliché.

Indeed, many writers took advantage of the prominence of heart disease to give their work relevancy, scope, and dramatic power. The Victorian sensationalist novelist Wilkie Collins, who was as up-to-date on medical matters as any fiction writer of his era, used heart disease as an important plot element in his greatest novel, *The Woman in White* (1860), when the heroine, Anne Catherick, dies of cardiac failure while being pursued by a dastardly nobleman. In George Eliot's *Middlemarch* (1874), a stingy justice of the peace, Arthur Brooke, is able to retire from a misguided political campaign by bogusly claiming to suffer from a heart ailment, while the pompous, dried-up clergyman Casaubon dies of heart failure after a short marriage to the novel's heroine, Dorothea. Indeed, cardiac disease appears with such frequency in the era's fiction that it is tempting to compare it to the plot device of the serial killer in our own time: a radically dramatic means of quickly upping the ante in any story.

One of the most striking examples of the morbid heart in the era's literature can be seen in the character of Arthur Dimmesdale in Nathaniel Hawthorne's *The Scarlet Letter* (1850). Although the novel's action takes place in Salem, Massachusetts, in the middle of the seventeenth century, the view of the heart that it presents is very much in keeping with the time of its publication. Dimmesdale may wear the clothes of a Puritan clergyman, but his ailments are very much those of a Victorian poet.

Like Casaubon and Anne Catherick, Dimmesdale shows clear signs of cardiac disease: "He was often observed, on any slight alarm or other sudden accident, to put his hand over his heart, with first a flush and then a paleness, indicative of pain." The source of Dimmesdale's heart trouble is his affair with Hester Prynne, which he has somehow managed to keep hidden from his parishioners. This deception, however, is exacting a terrible toll on the otherwise healthy young man. It is the "one morbid spot . . . infecting his heart's entire substance."

In an attempt to treat this infection himself, Dimmesdale undertakes a course of bodily mortification, "practices more in accordance with the old, corrupted faith of Rome than with the better light of the church in which he had been born and bred. In Mr. Dimmesdale's secret closet, under lock and key, there was a bloody scourge"—a whip. But even these extreme measures are of no use. The only likely remedy for his failing heart is reconciliation with Prynne, as evidenced by his brief period of good health after the couple meet in the woods and decide to flee Boston together.

Dimmesdale's condition is made all the worse by his choice of physician—Roger Chillingworth, Prynne's embittered husband. Recently arrived in Boston after being held captive by Native

Americans, Chillingworth keeps his identity hidden after learning of his wife's adultery. The "leech," as he is called, is not even a real doctor; he only pretends to be one so he can learn more about Prynne's betrayal. Like Dimmesdale, he looks like a man of the seventeenth century, but is very much of Hawthorne's era. Although he is well versed in the medical practices of his time, he also has a nineteenth-century belief in the connection between a person's emotional makeup and cardiac morbidity. "Wherever there is a heart and an intellect," he claims at one point, "the diseases of the physical frame are tinged with the peculiarities of these."

Thus, sensing that his patient's heart trouble stems from a spiritual wound, he starts probing the clergyman ever more deeply. Finally, after he moves in with Dimmesdale to better treat him, he glimpses evidence of his adultery in the form of an unspecified stigma on the flesh over his housemate's heart. Armed with this knowledge, the leech begins to torment his rival. Chillingworth proves very well versed in the proper ways to exploit a weak heart for his own twisted ends.

> The victim was for ever on the rack; it needed only to know the spring that controlled the engine;—and the physician knew it well! Would he startle him with sudden fear? As at the waving of a magician's wand, uprose a grisly phantom,—uprose a thousand phantoms,—in many shapes, of death, or more awful shame, all flocking round about the clergyman, and pointing with their fingers at his breast!

Chillingworth is pictured not only as an inquisitor, but also as something of a grave robber: "He now dug into the poor clergyman's heart, like a miner searching for gold; or, rather, like a sexton delving into a grave, possibly in quest of a jewel that had been buried on the dead man's bosom, but likely to find nothing save mortality and corruption." He even taunts his patient when he tells Dimmesdale that he has found strange, hideous plants growing on the unmarked grave of some poor sinner. The cunning cuckold suggests that the black weeds "have taken upon themselves to keep him in remembrance. They grew out of his heart, and typify, it may be, some hideous secret that was buried with him, and which he had done better to confess during his lifetime." Dimmesdale, clearly stung by these barbs, contradicts his physician, telling him that the unburdening of human hearts on Judgment Day creates not a tangle of weeds, but rather a beautiful flowering of joy and deliverance. "And I conceive, moreover, that the hearts holding such miserable secrets as you speak of will yield them up, at that last day, not with reluctance, but with a joy unutterable."

This point of view highlights another aspect of Dimmesdale's character that is wholly in keeping with the era's view of the heart: his compassion. The same sensitivity that causes him to suffer also endows him with a unique sensitivity to the pain of his fellow human beings. The Puritan shows himself to have something in common with Rousseau and Wordsworth: an acutely feeling heart. "His intellectual gifts, his moral perceptions, his power of experiencing and communicating emotion," the reader is told, "were kept in a state of preternatural activity by the prick and anguish of his daily life." Dimmesdale is said to belong to that class of preachers capable of "addressing the whole human brotherhood in the heart's native language." He possesses "sympathies so intimate with the sinful brotherhood of mankind; so that his heart vibrated in unison with theirs, and received their pain into itself, and sent its own throb of pain through a thousand other hearts, in gushes of sad, persuasive eloquence."

In the end, Dimmesdale does not flee with Prynne, but instead publicly unburdens himself of his sins before his flock, a confession that culminates in his literally baring his chest to reveal that mysterious stigma. Hawthorne never makes clear the exact nature of this mark. Most in the crowd report seeing a scarlet letter that mirrors the one Prynne has been forced to wear, though the author leaves open the possibility that the vision of the blot was the product of some sort of mass delusion. (The story is set, after all, in the Salem where just a half century earlier witch trials were held, one of whose presiding judges was in fact related to Hawthorne.) The source of the stigma, if it exists, also remains unclear. Is it self-inflicted? Does it have a supernatural origin, either infernal or divine? Or is it simply another outward manifestation of the patient's inner turmoil and compassion? Whatever the truth is about what is *on* Dimmesdale's chest, Hawthorne leaves us in no doubt about what happens *within* it: his tortured heart finally fails. After confessing his adultery, he drops dead in front of his congregation. In an era rife with literary infarctions, the tormented preacher's spectacular death is surely the most memorable.

WITH this epidemic of anxiety came the demand for new and better treatments. Drugs began to be used with increasing frequency and sophistication. In the early 1780s an English physician and chemist named William Withering learned that a local folk herbalist in his native Shropshire had used a concoction of native plants to cure a patient of congestive heart failure, or what was then known as dropsy. After analyzing the old woman's medicine, Withering concluded that the active ingredient was foxglove, a plant that takes its name from its flowers' resemblance to gloved fingers. Convinced that he had stumbled upon an effective tool in treating heart disease, he undertook a long study of the plant, reporting his findings in the seminal book *An Account of the Foxglove and Some of Its Medical Uses* (1785). It is the first systematic description of how a medicine can be used for therapeutic purposes.

Doctors and, perhaps more important, folk healers had long known that this plant had special qualities. What Withering determined was that foxglove acted as a diuretic, reducing the edema (or swelling) of heart congestion. Years of extensive subsequent experimentation established that the plant's active ingredient, digitalis, was also effective in reducing elevated heart rates and arrhythmia. By the nineteenth century, it was the drug of choice among doctors faced with a growing epidemic of heart disease.

Another medication that came into use during this time was salicin, an anti-inflammatory agent derived from willow trees. It had actually been employed as a pain reliever for thousands of years, and Hippocrates refers to the willow's bark and leaves as being effective against fevers. In the late 1820s, a number of European doctors, working separately, were able to isolate this bitter yellowish crystal. It was soon converted into the far more palatable salicylic acid, which became useful in treating a number of heart ailments, notably rheumatic fever. Perhaps its most important role, however, was as a direct precursor to aspirin, whose effectiveness as an anticoagulant to help prevent heart attacks and strokes is now well known.

This period also saw the development of nitroglycerin. In 1867 a Scottish researcher, Sir Thomas Lauder Brunton, gave amyl nitrate to a patient to relieve angina. It worked like a charm, though its side effects, which included a splitting headache, were daunting. So he tried a related substance, nitroglycerin, and found that it was even more effective (and with less dramatic headaches). Both drugs are vasodilators—they work by relaxing the blood vessels, particularly the large arteries. To this day, nitroglycerin continues to be used by patients to lessen the effects of angina.

Although drug treatments were showing promise in alleviating some symptoms of heart disease, there was precious little that doctors of this era could do to *treat* the root causes of these ailments. When they did intervene, their methods often had as much in common with Victor Frankenstein's as they did with those of today's surgeons and cardiologists. One of the most common interventional means of treating heart disease was *medical electricity*, the practice of sending electric currents into the body. The device used to generate this voltage, the galvanic cell, was a bulky, bipolar contraption—think of your car's battery—that used chemical reactions to produce positive and negative currents. (Jean Aldini used an early version of this on his hanged men and Poe referenced its use in his "The Premature Burial.") Dials could adjust the flow according to a set of guidelines that now seem as arcane as a necromancer's incantations.

One of the more detailed contemporary accounts of the use of the galvanic cell comes from *Medical Electricity: A Manual for Students*, published in 1873 by William White, a professor at the

New York Medical College for Women. His therapy for the combination of cardiac hypertrophy and palpitation, which he claimed afflicts people "of delicate constitution, lax fibre, and nervous excitability," as well as those prone to "any violent mental emotion [or] too free use of intoxicating drinks," is characteristic of the era's treatment for these conditions:

> If we find that the disease results from any local cause, we must treat according to positive or negative condition of the organ or part affected. If there is enlargement of the organ, ossification of its valves, obstruction, or inflammation, treat with a gentle [primary current] positive over the region of the heart, and negative on the spine, a little below the positive. Treat from five to eight minutes, and finish with general tonic treatment, avoiding the heart, and using [a secondary current], with patient seated on the positive. If there is *contraction* of the heart, treat over the heart, with gentle [primary current], negative over the organ, and positive over the cardiac plexus on the spine.

And then, once fully cooked, garnish the patient with herbs and serve with new potatoes, one is tempted to add. Although it is difficult for us to encounter such a passage without shaking our heads at the unintended harm such treatments might have caused, this aggressive application of electricity to the heart would almost certainly have had a hit-or-miss effect, producing favorable results in at least a few cases without the practitioner really understanding the physiology of what he had just done.

It is easy, of course, to mock an outdated therapy from the

safe removal of a century and a half of technological progress, though many accounts from the time show doctors laboring heroically with techniques that even now seem surprisingly modern. In a report on an inquest published in the June 1854 issue of the *Western Journal of Medicine and Surgery*, Monsieur Richard, a surgeon at a leading Paris hospital, demonstrated not only a potentially more beneficial application of the galvanic cell than White's meddlesome zapping, but also the state of cardiac resuscitation at midcentury. After Richard's patient, who was having a polyp removed from her uterus, went into arrest while being administered chloroform, the surgeon summoned the era's equivalent of a crash team to assist him.

> "I then produced artificial respiration by making rhythmical pressure on the chest. At the same time the assistants resorted to blows on the thighs, the calves of the legs, the arms, and friction on the face. Several times we titillated the top of the larynx. All these efforts were unavailing. On suspending the artificial respiration there were two or three long inspirations, after long intervals, and then cessation. Coldness became general but the beating of the heart was still heard, although feeble, by one of the *internes*. These efforts were continued for ten minutes. My two colleagues ... entered just as I had opened the trachea. I practiced insufflation through the orifice while some one continued the artificial respiration. M. Herard introduced into the precordial region two needles communicating with a powerful galvanic battery. The thoracic muscles were violently

convulsed, but no pulse was developed. At length we ceased further exertions after half an hour's duration of this terrible contest."

With the discovery of the heart's morbidity came a healthy amount of bald-faced quackery and hucksterism, with so-called doctors seeking to cash in on the newly minted public anxieties. One of the more notable instances was J. C. Boyd's Miniature Galvanic Battery, offered to Americans in 1879 for the low, low price of fifty cents (about $13 today). Its motto said it all: "The Blood Is the Life!" The item was a medallion—some rival versions even bore a heart shape in the center. Worn around the neck so that it rested over the solar plexus, it allegedly used the natural humidity of the skin to create a beneficial flow of electricity throughout the circulatory system. Professor Boyd promised that "nearly all diseases [were] effectually cured" by the production of the correct dose of electricity in the heart. The widely dispersed flyers for the product did not specify whether there was a money-back guarantee for those who dropped dead of ventricular fibrillation while sporting the bauble.

Given the growing awareness of the physical heart and its diseases, it is not surprising that the shape of the heart began to loom larger in the general consciousness during this time. The origin of the ♥ symbol itself is shrouded in mystery. It certainly predates even the most rudimentary scientific attempts to investigate the cardiac muscle—almost as if our collective unconscious understood the heart's value long before our higher faculties were able to chart its operation. Although some scholars maintain that the symbol is a crude representation of the vulva, others believe it to be an artist's attempt to render the tripartite heart described by Aristotle and his followers. It most likely has its roots, literally, in Greek Libya, where the heart-shaped (and extinct) silphium root was used as an aphrodisiac in the seventh century BC.

Whatever its origin, the shape underwent a radical growth in popularity during this time. In religion the image of the sacred heart, so aggressively banished by Protestant reformers, became central in Catholic iconography. Although, as we have seen, Jesus's heart had been an important part of worship since the Dark Ages, the heart logo was officially recognized only in the nineteenth century, when the church finally accepted the visions of the French visionary nun Saint Margaret Mary Alacoque (1647–1690) as authentic. A delicate woman given to intense bouts of self-mortification, Alacoque would regularly describe visions of Christ displaying his heart to her. In addition to being flaming and bleeding, the heart was also bound by a crown of thorns that looked like a strand of barbed wire. This image, whether appearing on Christ's chest or standing alone, eventually became the icon of the Society of the Sacred Heart, an exclusively female educational congregation that is often seen as a sister organization to the Jesuits. Although the image of this pierced heart is prevalent in paintings, perhaps its most popular use is on a scapular, two cloth badges connected by strings that run over the shoulders, and are situated over the chest and between the shoulder blades.

A more secular popularization of the heart icon was to be found with the valentine greeting, which came into widespread use during the Victorian era, particularly in England. Like the heart shape itself, the history of the valentine is murky. Saint Valentine of

Rome was purportedly an early Christian priest who was martyred by the Romans on February 14, 270, after illegally conducting marriages among the faithful, though it is more likely that his execution was for the more generalized offense of simply being a Christian. Whatever the reason, that mid-February date came to be associated with the commemoration of romantic love. The first known valentine was sent in 1415 by Charles, the Duke of Orléans, to his wife as he was being held prisoner in the Tower of London. The practice gradually grew in popularity, with couples exchanging handwritten notes and tokens. By the mid-nineteenth century, valentine cards were being factory-produced in both England and the United States.

The mass distribution of the heart image contributed to a phenomenon that began in the Victorian period but did not become full-blown until the twentieth century: the use of the heart symbol as both a commodity and an object of kitsch. As the medical revolution robbed the heart of its mystery and the Industrial Revolution ramped up the means of production, the heart became a cog in the commercial machine, a precursor to the smiley face. Even the veneration of the physical heart changed from the profoundly sacred worship of the Middle Ages to an empty ritual that had more in common with pop culture than with the adoration of the saints. Both Lord Byron and Frédéric Chopin—the Mick Jagger and John Lennon of their respective eras—had their hearts separated from their corpses and given special treatment. This process reached a black-comedic conclusion in the (perhaps false) legend of author Thomas Hardy's heart, which was allegedly eaten by the family dog after it was presented to his widow.

DESPITE the growing familiarity of the heart both as a commercial image and as an object of scientific study, the nineteenth century ended with cardiology in crisis. The era's great breakthroughs in anatomy, physiology, and nosology (the classification of diseases) served only to emphasize the ongoing, fundamental powerlessness of doctors to treat an ever-growing list of heart ailments. Physicians knew better than ever what could go wrong with the heart; they were just at a loss as to what to do about it. The development of surgical methods was clearly needed, though this was deemed by many to be not merely difficult, but impossible. As the doctor (Galen, in this case) had ordered, *noli tangere cordus*—"do not touch the heart." It was a sentiment echoed by the great Austrian surgeon Theodor Billroth, who claimed in the 1880s that "an operation on the heart would be a prostitution of surgery." As late as 1896, the English surgeon Sir Stephen Paget claimed that "surgery of the heart has probably reached the limits set by Nature to all surgery: no new method, and no new discovery, can overcome the natural difficulties that attend a wound of the heart." Drugs, electricity, and bed rest were still the treatments of choice. For the well-to-do, a stay at a luxurious Alpine spa might be prescribed. And yet, even this was of no real help. For all the mineral waters imbibed, bracing air gulped, and fortifying walks taken, it was often simply a matter of making the inevitable more palatable for those suffering from heart disease. Another revolution was needed, one as bold as that effected by the anatomists of the sixteenth century. It was time to open up the chest and start operating on the living, beating heart.

CHAPTER 5

Current Heart

Auguste-Viktoria Hospital, Eberswalde, Germany, 1929

He will not lose his nerve this time. Last week he was able to push the catheter only as far as his shoulder before abandoning the procedure. Of course, he'd had a sound clinical reason for the retreat: the tube had caught on the subclavian vein as it entered the vessel. He did not want to tear its inner lining. If he had started to bleed, there might have been no stopping it, especially since the only other person in attendance was Peter Romeis, a first-year surgical intern like himself. Despite his big talk, Romeis was only too happy when he decided to pull it back out. In fact, Romeis had insisted on it. He was clearly overwhelmed by the magnitude of what they were doing. And so Werner Forssmann had slowly withdrawn the catheter, trying to mask his bitter disappointment as he let his friend dress the wound on his elbow where he had inserted it.

Today, however, there will be no turning back. Just to be sure, he has decided on a new assistant: Gerda Ditzen. She is far more likely than Romeis to help him succeed. Plus, as a scrub nurse, she has access to sterile equipment. He had gone to her a few days earlier and told her of his dream. She had been hesitant at first, though he had soon won her over. He could see that she was excited by the prospect of being part of something that no one had ever tried before.

He knows that if he does not succeed today, he will never do it. After making sure that there are no surgical procedures taking place in the hospital's operating theaters, he approaches his accomplice at the busy nurses' station.

"Nurse Gerda," he says as casually as possible. "Please organize a set of instruments for a venesection and meet me in operating room number three."

Her cheeks redden, but the other nurses are ignoring them anyway. Venesections are routine. Luckily, no emergency cases have been brought in since he last checked. They will have the surgical wing all to themselves. As he waits for Gerda to arrive, he tries not to think about what his instructor, Richard Schneider, will say when he finds out what he has done. This is not only a serious breach of hospital policy. It is also a violation of one of the first rules taught to Forssmann when he was a medical student in Berlin: a doctor does not experiment on himself.

But rules are made to be broken, are they not? Just a few centuries ago, they had burned men at the stake for attempting these sorts of breakthroughs. What would they do to him? Slap his wrist? Put a letter in his file? Besides, if he is successful—*when* he is successful—no one will care about policies or rules. They will be too astonished by what he has accomplished. Because Forssmann, at the age of twenty-five and a doctor for less than a year, will be the first person ever to successfully introduce a catheter into a beating human heart.

It will work. He is certain of it. So certain, in fact, that he is willing to bet his life on it—literally. And it is not as if he is operating completely in the dark. Doctors had been performing cardiac catheterization through the jugular veins of horses for fifty years now. The dogs Forssmann himself has experimented on have all survived. Well, with the exception of that one mutt, but that had been his first attempt. He has practiced on cadavers as well—the catheter seemed to find its own way through the venous system. And it will be even easier when it is helped along by the flow of his blood. No, it is only superstition that has kept doctors from trying the procedure on humans.

Besides, he has already attempted to go through the normal channels. Schneider, the hospital's chief of surgery, encouraged him at first—as long as Forssmann confined his experiments to animals. And the intern could tell that the veteran doctor was intrigued by the notion of catheterizing a living human heart. But like Romeis, he lacked the nerve. Experimentation on people is strictly forbidden. The heart, Schneider claimed, could never withstand the shock of such a penetration. A foreign body would trigger an autonomic reflex action and cause a fatal attack. Or the catheter would shred the myocardial tissue and the patient would bleed to death. Of course, Schneider had absolutely no data to support this. Because no one had ever been brave enough to try it.

Until today.

It will be a huge leap forward. Not quite on a par with Harvey's accomplishments, but certainly as momentous as anything done by Hales or Marey. It will provide doctors with an entirely new way of diagnosing heart disease—and then treating it. Drugs will now be delivered directly into the diseased myocardial muscle. Surgeons might even be able to use the catheter to resuscitate hearts that have stopped beating. Forssmann himself will become a sensation. He will finally be able to leave this backwater for a big city hospital. They will ask him to demonstrate the procedure in London, Paris, New York.

Perhaps one day there might even be a trip to Stockholm.

Gerda arrives, breathless with excitement. She is, after all, only eighteen, a Silesian farm girl involved in something bigger than she'd ever dreamed possible when she first started her training. Her shaking hands rattle the sterile instruments a little. He gestures for her to place the cutdown tray on a table.

"Did you see anyone?"

She shakes her head.

"No Schneider?"

"No Schneider."

"Let us get to work then."

She hesitates.

"Doctor, wait."

"What is it?"

"I have been thinking. Do it to me."

"What?"

"The procedure. You are too important. If anything should happen . . . Do it to me. I'm just a nurse."

"Gerda . . . "

"Besides, it would make it easier, would it not?"

It is true. It would be much more straightforward to perform the procedure on her. Having both hands free would be useful. But the risk is too great. Not to her life—he is certain there will be no harm done to anyone today—but to his career. Operating on himself might be considered unethical. Doing it to her without his superiors' authorization, however, would be criminal.

"No," he says. "But I thank you for your offer. Now . . . "

He moves toward the instruments on the table, but suddenly she is standing in the way. Her eyes are alive with defiance.

"I cannot let you do this to yourself."

"Nurse, please . . . "

"If you don't do it to me, I will summon Schneider."

Forssmann can see that this is no empty threat. She might be a young country girl, but she also stares down death every day on the wards. And she'd had the courage to join him in this great adventure in the first place. Perhaps he has been naive in thinking that his authority would make her pliable. Perhaps he knows more about the physiology of the heart than he does about how it works in a young woman. He has to think of something. Quickly. She will ruin everything if she tells the surgical chief.

"Yes," he says, stalling for time, "you are right."

He can see the relief in her eyes. But also the dawning fear as she understands that she has just volunteered to have an apprentice surgeon insert a rubber tube into the right atrium of her one and only heart.

"Where do you want me?" she asks bravely. "Shall I sit in the chair?"

He looks around the room. And then he understands what he must do.

"Lie down," he says, gesturing to the operating table.

"Why?"

"The painkiller might make you faint."

After a moment's hesitation, she complies.

"And I'm going to need to strap you down."

Her eyes narrow with suspicion.

"Why?"

"Your muscles could go into spasm as I thread the catheter through. I cannot have you thrashing your arm around."

She nods, not happy about it but once again accepting his authority. He quickly fastens the leather straps around her wrists and ankles, trying not to notice how pretty they are.

"Will I feel it?" she asks, her voice tremulous.

"I will use an anesthetic on the arm, but there might be some discomfort . . . "

"No, Werner. In the heart. Will I feel it *in the heart*?"

"No," he says quietly.

Of course, he knows nothing of the sort. True, Harvey had said the heart was insensate. But he had never inserted a tube into its core. On the other hand, Forssmann is not really lying to the nurse. Gerda Ditzen will *not* feel it. Of that he is certain. Because he has no intention of doing this to her.

After checking the straps, he walks back to the table that holds the tray. He can sense her eyes on him as he sits in a chair and rolls up his left sleeve.

"What are you doing?" she asks.

"I am inserting a catheter into the right atrium of my heart, Nurse Ditzen."

"You tricked me!"

"For your own good."

"I'll scream."

"Then I will be forced to gag you."

She starts to respond but stops herself, understanding that she is helpless.

"If I drop dead," he says with a mirthless laugh, "*then* you may feel free to call for assistance."

He sets to work. He coats the inside of his elbow with iodine, then applies the local. After giving it a moment to work, he grasps a scalpel and makes an incision in the cephalic vein on the inside of the elbow, the one normally used for intravenous injections. He stanches the blood, thinking that this is where he could have used Gerda's help, if only she had not proved to be so stubborn.

He is ready for catheterization. He takes the rubber urethral catheter from the tray, smiling a little to himself at how different this use is from its intended purpose. But the smile soon disappears as he begins to thread it into the vein.

"Set me free," Gerda says as she sees him grimace. "I can help you. Please. I'll be good."

He is tempted, but it is too late. The great experiment has begun. He grits his teeth and continues to push the catheter deeper into his body. The tube moves easily through the upper arm. It is, after all, floating downstream. He tracks its progress in his imagination. It quickly reaches the shoulder—this is where it got caught last week. But today it passes smoothly into the subclavian and begins its descent toward the heart.

Suddenly there is a strange sensation in his upper chest. And then he coughs loudly.

"What was that?" Gerda asks in a panic. "What is happening?"

"I believe I just tickled my vagus nerve," he says with a soft laugh.

Checking the amount of tube he has threaded into his body, he calculates that it has now left the subclavian and is in the superior vena cava. He is getting close. He must be just outside the atrium. Poised at the heart's door. But he does not make the final push. Not yet. He wants to be able to watch this happen.

"All right," he says as he rises from the chair.

"What are you going to do now?"

He smiles at her.

"Take a photograph."

He takes a step toward the door.

"Werner—don't leave me here!"

She is right. That would be cruel. Besides, there is nothing she can do to stop him now. Gerda's eyes are on the tube projecting from his arm as he undoes her straps. Her anger has vanished, replaced by awe. He helps her to her feet. They face each other in silence for a moment.

"Are you really going ahead with this?" she asks.

He nods as casually as possible.

"We must go."

He has her phone Eva, the X-ray nurse, and tell her to meet them at her post in the basement. The old warhorse is there when they arrive. Her expression grows curious and then alarmed when she sees the tube poking from Forssmann's arm.

"What is this foolishness?" she asks. "Where is the patient?"

"I need you to take an X-ray of my heart."

"Your heart? Why?"

"Because I am about to insert a catheter into my right atrium."

Eva scowls to let them know that she does not find the joke

funny. But then her eyes travel from Forssmann, to Gerda, to the tube—and, finally, back to Forssmann.

"What have you done?"

"Something no one has ever done before."

"Is this true?" Eva asks Gerda.

"Yes," she says proudly. "It is true."

For a moment Forssmann thinks the older nurse is going to storm off to summon help. But then something new enters her expression: admiration. Admiration and awe. She moves efficiently as she fetches the bottle with the dye solution and injects the bright contrast liquid into the catheter. She then positions Forssmann behind the fluoroscope screen.

"Wait," he says. "I'll need to see it."

"There's a mirror in the cabinet," Eva says.

Gerda holds it up so he can watch himself work. He can see right away that his calculations were correct. The tip of the catheter is just a few centimeters from the right atrium. All it needs is a little push. And then history will be made.

Before he can do anything, however, the door bursts open and a breathless Romeis enters. He must have figured out what was happening after noticing the missing intern and nurses.

"Werner, you fool," he says. "You have to stop this."

"It's too late."

Romeis steps forward, looking as if he intends to yank the catheter right out of his colleague's arm. Forssmann raises a foot to give him a kick, but luckily Gerda quickly blocks Romeis's way.

"What are you going to do?" she says. "If you touch him you might kill him."

Romeis's shoulders slump. He knows he is defeated. He gives Forssmann a desperate look, then nods his head.

"Take a photograph before I start," Forssmann says to Eva.

She works the camera attached to the machine, exposing one of the large plates.

"Wish me luck," the young doctor says.

He takes a deep breath and gives the tube a firm push. The tip of the catheter enters the atrium. His heart has been penetrated. And yet he can feel nothing. Just as Harvey claimed.

"Have you really done it?" Romeis asks.

"I believe I have."

Forssmann holds perfectly still for a moment, unable to believe what he is seeing reflected in the mirror. And then his confidence wavers and he suddenly fears an explosion inside his ribs. His heart will understand what is happening. He will keel over in full arrest and enter the eternal blackness of death. But there is nothing. He feels as he always has. Better, perhaps. The excitement and tension energize him. He has done it. He is the first man in history to catheterize a beating heart. No one else has achieved this. Not Harvey. Not those sophisticated Frenchmen, Laënnec and Marey. Not those tough Scots, James Mackenzie and Lauder Brunton. But he—Werner Forssmann. His intubated heart starts to pound with excitement. Calm down, he tells himself. This is not the time for palpitation.

"Please take another photo."

Eva does as she is told.

"How does it feel?" Gerda asks.

"Like nothing at all," he answers.

He gives it another push. It travels even farther into the

heart. And still he feels nothing. He nods for Eva to take one more photograph.

"Will you enter the ventricle?" Romeis asks.

"I would, but the catheter is not long enough."

"So what happens now?"

Forssmann smiles at his friend.

"Get Schneider."

"He will probably dismiss you on the spot."

"Perhaps." Forssmann points at his x-rayed heart. "But he cannot dismiss *this*."

Schneider does not dismiss him on the spot, however. In fact, he is so impressed that he immediately recommends the intern for a position with the great Ferdinand Sauerbruch at the Charité Hospital in Berlin, the most sought-after internship in Germany. He also helps him publish a report of his work in the prestigious *Klinische Wochenschrift*—which turns out to be Forssmann's undoing. Once Sauerbruch learns that his new intern has been involved in self-experimentation, he sends him packing, telling him that a guinea pig will never be a heart surgeon.

Undaunted, Forssmann returns to Eberswalde, where he focuses on pure research. But then the tabloid press picks up on what he has done, making the young doctor seem like a carnival freak. He continues to work for the next few years, using his own body to prove conclusively that cardiac catheterization is safe and effective. It is rumored that he gives up only after he has subjected all of his accessible veins to cutdowns. It is more likely, however, that the disdain of his peers and the constant threat of disciplinary action finally drive him from cardiology.

He becomes a urologist and then, after serving in World

War II, he settles down to life as a country doctor who supports his wife and six children with a second job as a lumberjack. He would have died in obscurity had his paper not been read by two eminent American cardiologists a decade after its publication. They base their pioneering work in cardiac catheterization on that initial bold experiment. When the time comes for them to be awarded the Nobel Prize in Physiology or Medicine in 1956, Forssmann is honored as well. Although it is much later than he had counted on when he first opened his cephalic vein, his dream of traveling to Stockholm has finally come true.

IF the eighteenth and nineteenth centuries were the era in which doctors discovered the heart's morbidity, then the twentieth century was when they decided to do something about it. Once untouchable, the heart became fixable. Doctors were no longer content to play the role of detectives trying to interrogate the suspect after a crime. Rather, they began to intervene to stop the act in progress—or to prevent it from happening at all.

The century, particularly its latter half, also saw a revolution in the public's attitude toward the cardiac organ. People were no longer entirely at the mercy of the muscle in their chests. It was seen as being responsive to care. When protected with aerobic exercise and a heart-healthy diet, as well as screening programs, new medicines, and interventional surgery, the heart became a partner in health. Some even viewed it as a sort of training buddy. You could make it better. In 1910 the great Scottish cardiologist James Mackenzie famously claimed that "a heart is what a heart can do." A century later, we have come to believe that what a heart can do—and what we can do with it—is almost unlimited.

Perhaps inevitably, this same period turned out to be the least fertile of those that had passed when it came to strengthening the heart's role as a metaphor for what is most human. It began to lose its sublimity. While doctors and researchers were pumping out new ways of thinking about the physical heart, artists fibrillated in a chaos of commercialism and kitsch as they attempted to find fresh metaphorical representations for it. In the second half of the twentieth century, as our ability to mend and improve the cardiac muscle increased at an often dizzying pace, the heart icon became the stuff of advertising campaigns, motivational lectures, lifestyle choices, and valentine candy.

True, the heart remained the metaphor of choice in references to our emotional lives, even after our passions had been so decisively relocated to the mind. And yet this usage was becoming increasingly difficult to justify. In the era of the brain, which commenced with the publication of Freud's revolutionary work in the first years of the twentieth century, use of the word *heart* often became glib, a cliché. Ultimately it was little more than shorthand for qualities that the speaker, if pressed, probably would have admitted had nothing to do with the actual heart. When we say someone has a big heart, we are talking about a set of behavioral characteristics whose origin can be found in the frontal lobe of the brain. (Unless, of course, the subject suffers from cardiac hypertrophy, in which case medical attention should be sought.) Love and courage and conscience—we all know that these traits are housed at the other end of the jugular from the heart's once richly populated chambers. And yet the metaphor will not go away. Even though our brains tell us differently, we leave our hearts in San Francisco, wear them on our sleeves, speak straight from them, and allow them to be broken. We cannot seem to shake the habit of speaking these lines. We have learned them by heart.

Meanwhile, doctors pressed on, writing a whole new script. As the twentieth century advanced, they probed more deeply into the heart, mapping its terrain with ever-increasing precision. With the advent of open heart surgery in the middle of the century, doctors were able to cut through the sternum, or breastbone, to open the rib cage and look directly at the beating heart. They could hold it in their hands. They sliced it open, cleared its blockages, and replaced its parts. It possessed fewer and fewer secrets.

As a result of this increased transparency, the heart is now perceived in ways never imagined in earlier times. Red and yellow and gray, it feels like a flexed biceps. Its contraction—fierce, sudden, relentless—is one of the most vivid examples of pure life on the planet. The heartbeat is an astonishingly complex feat of bioengineering. Harvey was right. It is not a wellspring. It is a powerful pump—two pumps, in fact. The pump on the right side sends blood through the lungs, where it takes on oxygen. The pump on the left propels oxygenated blood to the body, where it fuels all of our activities.

Structurally, an even better analogy than a pump is to imagine the heart as a house. It possesses four rooms. The two atria, or upper chambers of the heart, are the receiving rooms. Blood returning to the heart collects in these areas. The right atrium receives blood returning from the body, and the left, blood from the lungs. The two ventricles are the lower chambers of the house. From them the blood exits the heart, the right ventricle sending it to the lungs and the left to the body.

The right and left parts of the heart are separated by an impenetrable wall, or septum. The rooms on either side of this wall are joined by doors—the valves. The mitral valve separates the left atrium from the left ventricle; the tricuspid valve does the same between the right side's chambers. The mitral valve is a thin, leathery structure composed of two leaflets that lie within a fibrous ring. It looks like a Catholic bishop's ceremonial headgear, or miter—hence the name. Long, thin cords that anchor these leaflets to small muscles in the ventricle prevent the leaflets from billowing back and leaking blood into the atrium when they slam shut. The tricuspid valve on the heart's right side normally has three leaflets. Both valves operate in a similar fashion. When they are closed, blood returning to the heart collects in the atria. When the ventricles on the other side of the doors finish contracting, the pressure within them falls sharply. This allows the valves to open. Blood from the atria rushes through the open doors and refills the ventricles. The ventricles then contract, causing pressure to rise sharply and slam the doors shut, thus preventing blood from leaking back into the atria.

Doors also control the exit of blood from the heart's ventricles. The aortic valve is situated at the point where blood leaves the left ventricle. It has three thin leaflets that lie within a ring of tissue. It looks a bit like the hood ornament of a Mercedes-Benz automobile. Its partner on the right side, the pulmonary valve, is a similar shape. Both are pushed open by the contraction of the ventricles. All four of the heart's valves have orifices of about 1.6 square inches (4 square centimeters) and are considered to be stenotic—tightly narrowed, which prevents the free flow of blood—when they are less than 0.4 square inch (1 square centimeter).

Like any modern house, this structure is serviced by a plumbing system: the three coronary arteries. Odd as it may seem to the layperson, the heart does not derive any nourishment from the vast amount of blood that travels through its chambers. It must feed itself. The muscle of the heart receives oxygen only via the coronary arteries. These are the first arteries to branch off from the aorta, the primary artery feeding the body. They lie on the surface of the heart like a network of clinging vines, eventually penetrating into the muscle and dividing into smaller and smaller branches as they power the dwelling that never sleeps.

This house is also serviced by electricity. Known as the conduction system, it is the wiring of the heart. The sinus node, a group of cells located in the wall of the right atrium, is the heart's pacemaker. Roughly every second, it spontaneously fires an electrical signal that tells the heart to beat. When we are exercising or excited, the brain tells the sinus node to speed up. This electrical signal travels to a switching station, the atrioventricular node, which is a bundle of cells that rests between the upper and lower chambers. It also moves through a series of wires known as the His-Purkinje system that coordinates signals to generate a well-synchronized beat. Disruptions in the electrical supply do not just lead to the equivalent of flickering lights and shorted-out appliances. They can cause sudden death.

Armed with the confidence that they finally understood the heart, doctors and scientists began to intervene in its affairs. The era of heart surgery is often seen as dawning in 1902, when

Scottish surgeon Sir Thomas Lauder Brunton, who had earlier pioneered the use of amyl nitrate to relieve angina, performed an operation to cure a patient of rheumatic mitral stenosis, the abnormal thickening of the mitral valve. The operation was a success. Unfortunately the patient was a cadaver, and Lauder Brunton never did attempt the procedure on the living. The challenges posed by operating on a beating heart were simply too immense. Cardiology had to wait for this breakthrough until 1923, when Elliott Carr Cutler of Boston opened a hole in the left atrium of a live patient with a small blade attached to his gloved index finger. Working blindly, he sliced through the patient's fused mitral valve. Still operating by sense of touch, he then cinched a purse-string suture to prevent bleeding. And all of this was done to a *beating heart*. Cutler's technique, however, proved to have a prohibitively high mortality rate, and the procedure was effectively abandoned until after World War II.

There was halting progress over the next three decades, with surgeons, often using hard-won knowledge from the battlefields of two world wars, focusing on treating the great vessels connecting to the heart instead of the muscle itself. The Nobel Prize–winning French surgeon Alexis Carrel proved particularly adept in this field, demonstrating that sections of the aorta's wall could be replaced using pieces taken from arteries or veins elsewhere in the body. To perfect his grafting skills, Carrel took sewing lessons from a lace maker, a reminder that surgery is as much about manual skill as it is about book learning.

A major obstacle remained, however, before surgeons like Carrel and Cutler could work effectively within the cardiac muscle itself. They needed to be able to stop the heart—and then start it again. Until this could happen, incursions into the working heart itself would be rare and largely ineffective. The frustration the era's doctors felt was immense. They had the anatomical knowledge; they had the surgical skill. What they did not have was the time. And even if they could stop the heart, brain death would occur before they would be able to perform the needed surgery and restart the organ. The surgeons of the first half of the twentieth century were like well-trained, highly motivated bomber pilots who were ordered to attack a deadly enemy, only to be told that they did not have enough fuel to reach their targets.

Given these restrictions, a handful of cardiovascular surgeons in the 1940s and 1950s performed nearly miraculous feats, particularly in one of the most emotionally jarring areas in all of medicine: the congenital heart defects that afflict and often kill so many babies. One of the gravest of these syndromes is tetralogy of Fallot, named after the French doctor Étienne-Louis Arthur Fallot, who gave the first full account of it in 1888. As the name suggests, it involves four basic defects. The first, and most critical, involves the connection between the right ventricle and the pulmonary artery. The area immediately beyond the pulmonary valve, called the outflow tract, is too narrow and restricts the blood flow to the lungs. The second is a hole in the wall between the ventricles. This causes the third defect, in the aorta. The aorta normally conducts the flow of oxygenated blood from the left ventricle to the body, but in those with tetralogy of Fallot, the aorta communicates with both ventricles, further diverting the flow of blood away from the lungs and also sending poorly oxygenated blood to the body. Finally, the right ventricle is thick and weak because of the abnormal amount of work it is forced to do to compensate for the other defects. This four-fronted break-down results in cyanosis or blue baby syndrome as well as fainting episodes known as tet spells. Poignantly, afflicted children learn that if they squat, they increase resistance in the arteries in their legs and blood is forced back into their lungs to increase its oxygenation, thereby preventing the spells. In the days before surgery, there was no treatment or cure for this syndrome. Many children born with it died young; patients were unlikely to live past forty.

In the 1940s two doctors at Johns Hopkins University—cardiologist Helen Taussig and surgeon Alfred Blalock—with the crucial assistance of an uncredentialed but brilliant African American laboratory worker named Vivien Thomas, came up with a remarkable strategy for combating tetralogy of Fallot. They devised what became known as the Blalock-Taussig shunt, a new passageway connecting the artery that normally supplies the arm with oxygenated blood to the pulmonary artery. This juncture was made at a point beyond the blockage that was preventing the pulmonary artery from carrying an adequate supply of blood from the heart to the lungs. This greatly augmented the amount of oxygenated blood circulating through the body. Blue babies turned pink in a matter of minutes. As ingenious as this procedure was, it still did not cure the defect. For that, the heart itself would have to be operated upon. Not even doctors as bold as the Johns Hopkins team were ready for that.

ANOTHER pioneer in the treatment of congenital heart defects was the Minnesota surgeon C. Walton Lillehei, one of the most daring practitioners in the history of his profession. The condition he combated was atrial septal defect, a simple congenital abnormality in which the wall between the heart's two upper chambers is not closed completely. As a result, when blood returns from the lungs to the left atrium, some of it shunts through this hole back into the right atrium before it can be pumped to the oxygen-hungry body. In many cases, patients tolerate the effects of this condition for their entire lives without showing symptoms. If, however, the hole is large and a sufficient amount of blood leaks back through it, there may be adverse consequences. Blood pressure in the lungs increases, lessening the amount of blood the lungs can oxygenate, ultimately causing backed-up blood to shunt from the right atrium *back* to the left. This reversal of flow is devastating because the shunted blood never travels through the lungs and does not become oxygenated. At this point patients experience cyanosis—they turn blue.

Lillehei's dilemma was that he knew how to patch the hole, but could not do so without stopping the heart. What he needed was a heart-lung machine, though that device was still in the process of being developed. And so, in 1954, he devised a process called controlled cross-circulation, in which he attached the circulatory system of a severely ill eleven-year-old boy to that of his father. Oxygen-depleted blood was routed from the boy's large veins into his father's circulatory system, where it was then pumped through the older man's heart and lungs before being returned, oxygenated, to the boy. In effect, the surgeon turned the older man into a heart-lung machine to keep his child alive. This gave Lillehei and his associates nineteen minutes to sew a pericardial patch over the defect in the stilled heart, a relative eternity for surgeons who had previously been limited to the three minutes it takes the oxygen-deprived brain to begin to die. It was an amazingly innovative and daring solution. Over the next year, Lillehei operated on forty-five patients—almost all of them babies—with defects that would otherwise have been irreparable. Ultimately, the practice of cross-circulation had to be abandoned because of the risk it posed to both donor and child.

The problem faced by surgeons like Blalock and Lillehei had seemed near solving in 1953, when the heart-lung machine was first used for surgery in a human. Twenty-three years in the making, it was the product of Philadelphia surgeon John Heysham Gibbon in conjunction with a team of IBM engineers. Simply put, the bulky contraption took over the functions of the heart and lungs, removing depleted blood from the body and pumping it full of oxygen before returning it. Using his own technology, Gibbon performed what is considered the first complete open heart surgery—on an eighteen-year-old woman—keeping the patient alive on the pump for a half hour while he repaired a hole in the septum between her two atria. Although that operation was a success, further efforts were not so fortunate, causing Gibbon eventually to abandon surgery altogether. (This rocky start also explains why Lillehei did not have access to the technology at the time of his first operations.) Despite these setbacks, the era of open heart surgery had begun. The heart was literally open for business. Both doctors and patients could now think about the heart in ways that had been the stuff of science fiction just a few decades earlier. The heart could be stopped, cut into, altered, stitched up—and then started again. The next fifty years would see an explosion of techniques in the field of interventional cardiology that made the previous four thousand pale in comparison.

The greatest advances during this half century involved the treatment of the organ's most common and deadly ailment: coronary artery disease. When we think about our own hearts, it probably is not as the seat of our conscience, or the house of our god, or a metaphysical transmitter that allows us to send private messages to our one true love. Rather, we think about plaque—specifically, whether or not it is building up in our coronary arteries. And, more urgently, what we can do to get rid of this deadly substance before it kills us.

Although researchers had long suspected that lesions and blockages in the coronary arteries played a major role in heart disease, it was not until 1912, when Chicago physician James Herrick published an article titled "Clinical Features of Sudden Obstruction of the Coronary Arteries," that someone explicitly proposed that a heart attack resulted from a blood clot that had formed in a coronary artery. Although this is one of the most important observations in twentieth-century cardiology, it went largely unnoticed for almost seventy years, until Marcus DeWood of Spokane, Washington, reported threading catheters into the hearts of patients who had begun having heart attack symptoms less than twenty-four hours before. DeWood infused a contrast agent into the coronary vessels that was visible on X-rays. The results of this procedure, called coronary angiography, confirmed Herrick's hypothesis.

In the decades between Herrick and DeWood, many doctors possessed a limited understanding that there was a relationship between diseased coronary arteries and heart attacks. What they did not grasp was the nature of the connection. There was, therefore, little they could do to stop the coming heart attack. Treatment consisted of morphine, bed rest, and crossed fingers, which might alleviate symptoms but did nothing to cure the underlying condition. But in nearly every case that DeWood and others examined during their research with catheters, a clot had formed to block the artery, starving the heart muscle of oxygen. The deprived heart tissue *infarcted*—it died. If the infarct was severe enough, the heart stopped. It was a distant but powerful echo of the Egyptian metaphor of *metu*, in which the blood vessels were imagined as the Nile's vast tributary system. Here, detritus that floated downstream and formed a dam would soon block the tributary entirely. The resulting drought could be deadly.

At long last, doctors knew what caused that most terrifying of bodily events—the heart attack. They could now explain the phenomenon that the Greeks called lightning in the chest, the Victorians termed cardiac apoplexy, and we know by the more prosaic name of acute coronary syndrome (ACS). More important, they understood what had to be done in order to stop these attacks from happening. The dam had to be broken—or, even better, prevented from forming in the first place. The first weapon deployed was a blood thinner that had long been in the medicine cabinet: aspirin. Patented by Bayer in 1899, it blocks platelets, the circulating cell fragments that stick together to form a clot. A number of other anticlotting medications were subsequently developed, including heparin, a relatively crude drug made by boiling pig intestines, that is still in use today. Another drug commonly used to prevent clotting, bivalirudin, is a derivative of the blood thinner in leech saliva, suggesting that the doctors of antiquity may have been on to something all along.

But these drugs do not dissolve an already formed clot. This can be done only by drugs that activate plasmin, an enzyme that breaks up the congealed mass. Among the first of these drugs was a substance distilled from the urine of postmenopausal nuns in a process that had originally been used to create a fertility drug. The most widely used medication of this type now is tPA (tissue plasminogen activator), which put the first giant biotechnology firm, Genentech, on the map in the early 1980s. In a field of inquiry where progress was once measured in centuries, the speed of these developments was breathtaking.

The problem with these drugs was their side effects—not surprisingly, blood thinners could cause serious bleeding. So researchers began looking at ways to stop clots from forming in the first place. To do this, they had to identify what caused them. Cholesterol was soon identified as public enemy number one. The soft, waxy, fat-like substance had been associated with these catastrophic clots since the early part of the twentieth century, when Russian researchers fed rabbits diets of pure cholesterol and then watched them die of heart attacks. As the century wore on, cholesterol was increasingly implicated in coronary artery disease. It was not until the 1980s, however, that scientists truly understood how it worked.

The good twin–evil twin story line popular in countless television shows and pulp novels is perhaps the best way to visualize the process uncovered by researchers. There was "bad" cholesterol—LDL (low-density lipoprotein), which forms plaque—and "good" cholesterol—HDL (high-density lipopro-tein), which helps carry cholesterol away from its hiding place in the arteries to be broken down in the liver. In fact, however, no cholesterol is bad per se—it is essential in the formation of cell membranes. As is so often the case, the problem is a matter of degree. When too much cholesterol is present, the body tends to deposit it in the walls of blood vessels, leading to a disease known as atherosclerosis that may eventually cause the artery to narrow and close. This wreaks havoc, causing not only heart attacks but also strokes. One of the most intensive research-and-development campaigns in medical history resulted in statin drugs, which regulate the production of bad cholesterol by blocking an enzyme essential in its creation. Starting with the work of two Japanese researchers in the early 1970s, an increasingly wide array of scientists and a number of pharmaceutical companies labored to create a drug that would help keep arteries clear of the plaque that kills and disables so many. (Drugs promoting the creation of the good kind are proving more elusive, although one strategy—which involves looking more closely at populations of people in northern Japan who lack an enzyme that converts good cholesterol into bad cholesterol—suggests that such a therapy might be possible before long.)

Since the 1970s, the exact process by which plaque forms has come to be better understood, and the picture that's been painted is both more complex and more sinister than initially thought. It was not, as first believed, a matter of cholesterol sim-ply sticking to the inner walls of the arteries like gunk in a drain-pipe. Rather, inflammation occurs when cholesterol is deposited *beneath* the artery's lining. This cholesterol attracts immune cells that see it as a foreign invader and try to engulf it, much as in the classic Pac-Man video game. The subsumed cholesterol turns crystalline, attracting calcium, which hardens on top of it. And plaque is born.

This plaque is like a volcano. If the crystallized cholesterol erupts through its calcium cap on the artery's wall, it attracts platelets in the passing blood that quickly cause a clot. If the clot is big enough, blood flow to a portion of the heart muscle stops, causing an infarct. It was just such a volcanic eruption that caused one of the most newsworthy cardiac arrests of our time, when a fifty-two-year-old jogger dropped dead in Vermont in the summer of 1984. Although a middle-aged American male dying of a heart attack is by no means an extraordinary occurrence, the fact that the man was Jim Fixx made the story front-page news. At the time, Fixx was perhaps the best-known figure in America's burgeoning fitness cult, author of the runaway bestseller *The Complete Book of Running*. Slender, healthy, and boyishly enthusiastic, he was one of the first public figures to preach the beneficial effects of jogging on the heart. The fact that he was felled by a myocardial infarction was almost incomprehensible. It also gave ammunition to those who still thought that the heart was more an instrument of cruel fate than a muscle capable of being strengthened.

And then people started looking more deeply into Fixx's life. He had been obese as a young man and had smoked two packs of cigarettes a day until his conversion to a life of fitness at the age of thirty-five. His autopsy revealed severe blockages in his coronary arteries, including one that was 95 percent closed and another with an 85 percent occlusion. According to reports from family and friends, Fixx did not have a regular physician and went for checkups only rarely. Perhaps most significant, he had a troubling family history of heart disease—his father had suffered his first heart attack at thirty-five and died of his second at age forty-three. And so Fixx was not such an unlikely candidate for a sudden, fatal heart attack after all. His death did not mean that exercise is futile. It just meant that it is not always enough.

Fixx's death highlights one of the most important changes in the way we now imagine the heart—our understanding that our hearts are generations in the making. Like characters in a Greek tragedy, they often carry their fatal flaws within them, written in the intricate code of our DNA. In medieval times the heart was often thought of as a book upon which the contents of the soul were written. In the past few decades, the heart has come to be seen as a very different sort of text, one whose past is but prologue to its owner's future health.

The ideal reader of such a book is a doctor. Whereas the Egyptian and Christian hearts of ancient times had to wait until Judgment Day to testify about their owners, the modern heart can give evidence while it still beats. A good cardiologist is often relatively certain that a patient does or does not have coronary artery disease after just a brief discussion. Age is crucial—it is uncommon for patients under forty to have symptomatic coronary artery disease. Gender is also important, since men develop the disease at an earlier age than women do (although more women now die from cardiovascular disease than from all forms of cancer). And then, as Fixx's case reminds us, there is family history. If a parent or sibling has any kind of coronary artery disease before the age of sixty, this foreshadows trouble.

The heart itself speaks in ways the patient cannot. Doctors have been trying to read blood pressure since Stephen Hales impaled those luckless horses, though it was not until the early part of the twentieth century that physicians began describing the damaging effects of hypertension, which can harden blood vessels and lead to strokes and heart attacks as well as damage the kidneys. Now anyone can learn his or her ideal blood pressure, as well as the target numbers for cholesterol and heart rate. We have come to think of our hearts in terms of a series of measurements that can be compared to an ideal. Just as da Vinci's perfectly delineated human form, Vitruvian Man, can make every male feel inadequate, so too can a doctor's scrawled target levels on a drug-company notepad show us the dimensions of a model heart that will be tough to live up to.

Often the heart speaks through pain. Angina, a constricting discomfort in the chest (and sometimes the arm and neck), has been recognized and puzzled over for centuries, but it was only in the early 1900s that doctors began to suspect that it was caused by something going wrong in the coronary arteries. Now we know that this very particular mix of symptoms is perhaps the main indicator of coronary artery disease. We are taught to listen to our hearts, not just as a way of connecting with our deepest emotions or our deity, but also as a means of understanding the state of our physical health.

Anyone who has ever suffered from angina (or been close to someone who has) will know its symptoms well. More than any other condition, angina requires the sufferer to become a poet.

Shakespeare and Keats labored to find apt metaphorical means for describing the heart; the modern cardiac patient becomes their heir as he strives to tell his doctor about his myocardial muscle. Adjectives and metaphors become all-important. "Burning" and "squeezing" sensations are indicators of possible coronary disease, as are "pressure," "tightness," and "heaviness." A feeling of having a "band or rope around my chest" is a serious warning, while "a weight or elephant on my chest" is a dead giveaway that the heart is in trouble. The discomfort often occurs with physical activity, although the symptoms subside within a few minutes with rest. Angina may radiate pain to the jaw or left arm and can cause shortness of breath, nausea, or sweating. Other metaphors and descriptions point to other problems. "Sharp," "stabbing," "knifelike" pain; "pins and needles"; and pain that is reproducible when the patient is touched or that changes with position—these are seldom angina. Pain that lasts a few seconds or all day usually does not qualify. Women more commonly have atypical symptoms—fatigue, shortness of breath, nausea, or light-headedness—compared to men.

Often, the doctor deliberately puts the heart under duress to force it to speak. It is said that in the 1950s, Paul Dudley White, President Dwight Eisenhower's cardiologist, walked heart patients from his basement office to the third-floor operating room at Massachusetts General Hospital, taking their pulses and observing them closely as they ascended. If they looked all right upon arrival, he concluded that they were unlikely to have a heart attack during the operation. A more practical means of assessing the condition of a patient's heart had been developed in the 1930s, when a New York City cardiologist named Arthur Master constructed the first step-aerobics platform in his office. This gave way to the standardized treadmill protocol that is still used today for exercise stress tests, which puts the dash toward a diagnosis on a path that gets steeper and faster as the patient proceeds.

The problem with these stress tests was that they underdiagnosed heart disease, failing to pick up on underlying problems and suggesting that all was well. Like its owner, the heart proved capable of telling lies when put under pressure. So doctors began taking pictures. X-rays began to be used to look for heart disease soon after they were invented in 1895. Their limitations eventually became apparent, however, because they require a catheter to be inserted into the heart to introduce a contrast agent for the X-ray to pick up. Ultrasound, which became the technology of choice in the late 1960s and early 1970s (a period known as the sonic boom era), allowed doctors to visualize the heart as never before by converting sound waves reflected by the moving red blood cells into a color map. Gamma ray–emitting radioisotopes like thallium and technetium were soon being infused into the heart and its vessels through an intravenous drip. These particles would then emit radiation that was detectable by a gamma camera placed just outside the body to create radioactive atlases of the heart muscle. Newer high-tech modalities like PET (positron-emission tomography) and MRI (magnetic resonance imaging) create even finer pictures. Just a few decades after being invisible to the naked eye without cutting into the body, the heart has become almost fully illuminated.

The impact that this heightened awareness of coronary artery disease has had on the public's imagination over the past forty years has been as profound as Harvey's *De Motu* and the work of the morbid anatomists. When we think about the heart, we often think about how we can spend our time and money to maintain and improve its condition. We have been given agency over it. We *own* our hearts in a way our ancestors never did. The heart is becoming objectified and commodified in our minds; it has become big business. Some of the most compelling cardiac narratives now being produced are the work not of poets, painters, preachers, or playwrights, but rather advertising copywriters and film directors who are out to get us to spend our money to make our hearts stronger and more reliable.

Lipitor (atorvastatin) is the most commonly used statin. It is the drug of choice in the battle to reduce bad cholesterol. In the early part of the twenty-first century, its manufacturer, Pfizer, produced two television commercials that were broadcast extensively. Both spoke subtly yet powerfully to the hopes and anxieties that underlie our thinking about the heart. In "Skier" (2004), a spry, sixtyish man starts down a ski run on a rugged-looking slope. He is lean and vigorous. Initially he does not appear to have a care in the world. The mountain belongs to him. But then he realizes that he is being chased. As music that might have been the theme of a 1960s detective show sounds, a large placard bearing the phrase "255 Total Cholesterol" in red letters pursues him down the hill. (The Vitruvian ideal is less than 180.) No matter how skillfully he pivots and slides, the skier cannot shake off his pursuer. Finally he leaps a deep crevasse and skids to a halt. He looks around in triumph, thinking he is free.

But a female narrator informs us otherwise. "Sometimes all the right moves cannot give cholesterol the slip," she warns. Sure enough, as the skier turns away from the camera to continue his run, we see that the placard is stuck to his back like an alpine racer's number. All is not lost, however. The narrator tells us about Lipitor, which could help lower cholesterol when a low-fat diet and exercise alone fail. With the invocation of the product's name, the number affixed to the skier's back turns a healthy shade of green and lowers to a more acceptable 172. Our hero continues down a clear, serene slope.

In the other advertisement, "Diva" (2002), a limousine pulls up to the curb outside a large theater. A glamorous woman in late middle age emerges from the car and begins to make her way along the red carpet. All eyes are on her. The hand of a much younger man rests on her lower back—it is not clear whether he is a bodyguard, a lover, or both. There is applause; lightbulbs flash. As the diva progresses, admirable statistics appear in a box on-screen: "Height: 5' 9"." "Weight: 125." "Dress Size: 6." And then a discordant statistic appears, printed in white on a red background: "Cholesterol: 273." Suddenly the diva stumbles and falls, her plunge captured by the paparazzi. Stark white letters appear on a suddenly blackened screen: "High cholesterol doesn't care who you are." As the advantages of taking Lipitor are affirmed by the narrator, the diva is helped to her feet by her young escort. She is fine; she does not even seem to be embarrassed. The show will go on and, blessed by statins, she will be its star.

The advertisements clearly struck a nerve—Lipitor was the world's most popular branded pharmaceutical during both years. Worldwide sales would top $10 billion annually by the end of the decade. Despite their blandly upbeat veneer, both ads contain stark warnings of impending doom. Their subject is disease and death—and the avoidance thereof. These are not sick people we see on-screen. In fact, they are near-perfect specimens. Both move dynamically, confidently. They wear their years like finely tailored clothes. They are doing nothing wrong. And yet they are stalked by death. Both could be felled at any moment. The message is subtle but clear: your best efforts might not be good enough. Even if you live as admirably as these two, your heart might still strike you down. Only medicine can provide complete protection.

What is so cunning about these commercials is how they feed not so much on our desire for good health as on our fear of sudden death. Even the best of us are not in control of our destinies, they say; or, rather, we are in control of our fates only if we consume the product being advertised. And our awareness of death by heart disease has become so ingrained, so sophisticated, that it can now be suggested by imagery that appears at first glance to be benign, even comical.

This sophistication has been brought about in large part by an unprecedented revolution in the ability of doctors to treat the heart. More than ever, it is science, not art, that is changing the way we imagine the cardiac muscle. The messages encoded in the Lipitor commercials work because the modern consumer thinks about his or her heart in a whole new way. It has become a machine that can be tuned up by exercise, diet, and medicine and, if need be, repaired by experts.

This shift in imagination is not just the result of advances in diagnosis, as remarkable as they have proved. Nor is it due solely to new medications, as pervasive as drugs may be. It is also because of interventions—ranging from the seat-of-the-pants efforts of cardiac surgery's daring pioneers, to the sternum-splitting bravado of the open heart era that began in the 1950s, to the present day, when thousands of procedures are successfully performed daily, often with relatively little disruption to the life of the patient. Occasionally these interventions have made headlines and turned those performing them into international celebrities. More often, they have come about quietly, their effects rippling outward until they touch the lives of millions.

One of the most important developments in the history of cardiology came in Zurich in 1977, when a German cardiologist named Andreas Gruentzig, following in the footsteps of Werner Forssmann, inserted a balloon-tipped catheter into a patient's partially blocked left descending coronary artery after threading it through his arterial system. Once the balloon was positioned at the location of the plaque buildup, Gruentzig inflated it, compressing the plaque and expanding the vessel's inner cavity back to its optimal size. The artery remained widely open for a quarter century. Less than two decades after the primary cause of heart attacks had been discovered, a direct means of preventing them—or even reversing them while they were happening—had been developed.

Balloon angioplasty, as the process is known, has become an increasingly effective method for treating angina and coronary artery disease. Despite its revolutionary daring, Gruentzig's procedure was fairly crude. The artery-stretching balloon often damaged the vessel's intima, its inner layer, thereby activating the very cascade of clots it was meant to prevent. Heart attacks could occur, sometimes during the procedure itself. A lot of work had to be focused on finding the proper drug cocktail that would thin the blood enough to avoid clotting both during the procedure and afterward without causing an excessive risk of bleeding. Also, in about a third of the cases, scarring left behind by the barotrauma caused by the balloon would lead to later narrowing of the artery.

The solution was a stent, a slotted, stainless steel tube that looks like the spring in a ballpoint pen and is becoming a permanent part of an increasing number of human anatomies. (The device is believed to be named after a nineteenth-century English dentist, Charles Stent, who invented a scaffold to support facial tissues during early attempts at reconstructive surgery.) Since the stent received FDA approval in 1994, Gruentzig's balloon—in a more efficient version—has been used to deploy the device. After the balloon is deflated and withdrawn, the stent remains embedded in the wall of the artery and does not subsequently migrate. Early stents were fairly stiff and often could not be snaked through a winding or calcified artery to reach the site of a lesion. These were soon replaced by thinner, more flexible structures made of suppler alloys, such as chromium cobalt. Angioplasty became even more efficient with the use of immunosuppressant drugs—initially rapamycin, which is derived from a soil fungus found on Easter Island (also known as Rapa Nui), and paclitaxel, a cancer drug originally derived from the bark of the yew tree. Some stents are now manufactured to release these medications slowly. Ideally, a single layer of cells should grow over the metal struts so that, when blood flows by, it senses familiar cells rather than the stent's edges and is therefore less likely to clot.

Sometimes the blockage is so bad that stents are not enough. The situation resembles traffic accidents on high-volume roads. Drugs and stents are like the tow trucks that push damaged vehicles to the side of the road or tow them away. Sometimes, however, the pileup is too big to be dealt with this way—an eighteen-wheeler overturns and a bridge is damaged. The highway must be shut down and traffic redirected. In 1967 René Favaloro, a cardiovascular surgeon at the Cleveland Clinic, determined that a new route for this flow could be built when he successfully grafted a section of healthy vein from a patient's leg above and below the blocked section of coronary artery, bypassing it entirely. The coronary bypass, the most common form of open heart surgery, was born. In 2006, the American Heart Association reported, this procedure was performed on nearly a half-million patients. The unimaginable has become routine.

Often, more than one bypass is needed, leading to a comprehensive reconfiguration of the heart's blood supply routes. Healthy vessels are scavenged from throughout the body. The internal mammary artery, which supplies blood to the chest wall, is the vessel of choice because it is highly resistant to atherosclerosis. Veins from the leg usually supply the remainder, though some surgeons like to use a segment of the radial artery from the wrist.

One drawback with bypass surgery is that it involves cracking open the chest, bringing on the risk of infection and a prolonged recovery. Less traumatic forms of the operation are currently being developed; the routine is becoming even more routine. One such approach involves accessing the heart through the side of the chest, between the ribs, rather than by splitting the sternum. Surgeons are also feeling confident enough to perform some procedures "off-pump," such as sewing bypass grafts into isolated segments on the beating heart without the need for cardiac arrest and a heart-lung machine.

WITH this explosion of surgery came yet another evolution in how we imagine the heart. Christ's pierced and flaming heart, the amputated hearts in the works of Boccaccio and the Jacobean playwrights, the bootleg organ of Frankenstein's monster—these images of the besieged myocardial muscle gave way to the sutured heart of the modern postoperative everyman. This new thinking is given vivid representation by the British sculptor Damien Hirst, the enfant terrible of the contemporary art scene. Hirst is fascinated by the body—his most famous pieces include pickled animal carcasses and organs. In 2005, as part of his New Religion exhibit in London, which examined the conflict between science and religion, Hirst unveiled a sculpture called *The Sacred Heart of Jesus*. It consists of a bull's heart suspended in a formaldehyde solution within a Perspex vat. In a provocative commentary on Saint Margaret Mary Alacoque's vision, the heart cast in silver is pierced not by thorns or an arrow, but by hypodermic needles and a variety of different scalpels that cut into the muscle. It is diffi-cult not to feel an empathetic stab of pain when seeing the piece for the first time.

Hirst's work evokes a world in which the heart can still be used as a metaphor, though it has been drained of its old meanings and now struggles to accommodate new ones. The symbol has become utterly secularized. The sacred heart of Alacoque's visions was pierced by the sins of humankind; Hirst's is punctured by the imple-ments of human healers. It is no longer the repository of divine redemption and grace, but the object of earthly intervention and technology. The stigmata of medieval saints have given way to the surgical scars zippering the sternums of a generation of retirees.

The heart may endure as a symbol, Hirst seems to be saying, but it is a symbol that has been usurped by the scalpel.

THE coronary arteries were not the only area of the heart where doctors and surgeons felt emboldened to intervene once they could halt the pulse. They also made huge advances in treating the heart's valves. While malfunctioning valves inspired some of the earliest attempts at heart surgery, it was not until the 1960s that their treatment really took off. Replacement became the therapy of choice. The heart was not just fixable; its parts were becoming exchangeable.

The first luminary in the field of valve replacement was the aptly named cardiovascular surgeon Albert Starr. In 1960 he and Lowell Edwards, an engineer who also invented a hydraulic tree debarker and a fuel pump for airplanes, introduced what became the Starr-Edwards valve, the first artificial component to be implanted in the heart. It was the very picture of simplicity, being basically a plastic ball in a metal cage. The ball mimicked the working of a valve's leaflets by moving with the contraction of the heart and the pressure changes on either side of the valve. In the open position, it was pushed up in the cage and blood could flow through the ring and around the ball. In the closed position, the ball settled in the ring and blocked the flow. The problem was that this first artificial valve occasionally formed clots, which could break loose and cause strokes. It could also trap and pulverize red blood cells, causing anemia and jaundice.

And so another aptly named surgeon—the adventurous Viking Olov Björk—devised a tilting disk that bore a distinct, if not altogether reassuring, resemblance to a raised toilet seat. It was widely used in the 1970s, but a design flaw in one of the struts caused fractures, resulting in catastrophic valve failures that led to sudden death in 619 cases (out of eighty thousand implanted). This was enough to lead to a recall and eventually the product's withdrawal from the market in 1986. The difficulty, of course, is that when, for example, the hinge on a pickup tailgate is recalled for a similar flaw and fracture risk, it is much easier to replace. A valve in the heart requires considerably more time in the shop.

The experience hastened the next step in the technology—tissue valves, which are constructed from pig or cow pericardium, the leathery outer lining that covers the heart. Surgeons also began to use the entire valve from these animals, sewing it onto a metal wire strut. The preservative agent glutaraldehyde is used to stiffen the tissue, much as in tanning a hide. Another option is a homograft, which uses a human valve that has been explanted from a cadaver and cryopreserved with liquid nitrogen. None of these valves are living tissue; rather, they are a particularly expensive sort of leather. Thus, rejection is not an issue, as it is in transplants.

Valve repair has come a long way since Elliott Cutler poked his razor-tipped finger into the heart. Surgeons can now use less-invasive techniques. Some use a robotic device called the da Vinci Surgical System in honor of the artist who so memorably described valves' function more than five hundred years ago. The da Vinci consists of a stereoscopic video headset and joystick controller. Future generations of surgeons will benefit from having grown up with Nintendo's Wii, Microsoft's Xbox, and Sony's PlayStation, just as surely as Laënnec benefited from his flute lessons and Alexis Carrel from his sessions with that lace maker. The next ten years will also see interventional cardiologists getting in on the act, replacing valves in a beating heart using a catheter-based approach, much as they currently perform stent procedures. The day when open heart surgery will once again be rare might be approaching. This time, however, it will be not from a lack of confidence, but from an abundance of it.

Since the earliest days of scientific inquiry, doctors have understood that the heart's rhythm isn't just a sign of vitality; it can also reveal illness. Galen was obsessed with the pulse, isolating twenty-seven patterns of heartbeats and assigning them a dizzying array of names, from gazelle to worm. Although we now understand the heart's rhythms in ways that would have stupefied not just Galen but also doctors of a few generations ago, controlling the source of erratic heartbeats—arrhythmia—can still challenge even the most experienced cardiologist. They remain perhaps the most significant way in which the heart proves itself to be still a mystery.

As we have seen, most of the cells in the heart are capable of firing an electrical impulse. When they misfire—or short out—the result can be palpitation or even death. Palpitations, which usually result from a single premature heartbeat, can occur in a normal, healthy heart. Cases are on record of patients who experienced more than ten thousand premature contractions in the course of a day without being aware of them or needing treatment. Others are not so lucky. The condition can be a torment. While doctors refer to a palpitation as an *extra* beat, patients usually experience it as a *skipped* beat, followed by an unsettling thump. We now know that both descriptions are apt. What happens in palpitation is this: electrically, there is a premature beat. Often the heart has not yet filled with blood when it contracts, so there is nothing to be pumped. This is an "empty beat." It is not felt. This is the skip. There is then a pause before the next normal beat as the conduction system resets. During this pause the heart overfills with blood. Pressure and stretch sensors within the heart detect that and cause the next normal beat to be more forceful in order to deal with the increased volume of blood. This is the thump.

Palpitations have multiple causes. The most common are stress, inactivity, and stimulants ranging from cocaine and methamphetamine to coffee and decongestants. They can also occur without any obvious cause. The patient soon becomes aware of the heartbeat and awe turns into anxiety, compounding the problem. The doctor's daunting task becomes finding a way to reassure the patient that the palpitations will not cause a heart attack, a stroke, or sudden death. Such reassurance often makes the palpitations disappear, or at least become tolerable. The doctor can often calm a sufferer by recording the rhythm. While first-generation recording devices, which were deployed in the 1960s, were bulky and had a limited recording duration, two recent technologies have been much more useful, if a bit Orwellian. One involves a simple electrode that adheres to the chest and transmits the signal to a distant monitoring station. Patients can wear this device for up to one month. Another advance is a small recorder that can actually be inserted just beneath the skin and worn for up to one year.

Some heart-rhythm disturbances are more serious and require treatment. Such rhythms can either be too slow (bradycardic, or "brady") or too fast (tachycardic, or "tachy"). Old age is a common cause. The heart's conduction system, like the rest of the body, wears out over time. Its circuit breakers become frayed and weak. The sinus node may become unreliable, or the atrio-ventricular node, the switching station between the upper and lower chambers, may not consistently transmit signals.

Confronted with hearts that were bradycardic, doctors decided to pick up the beat. Artificial pacemakers were first conceived more than a century ago, when Swiss researchers converted the fibrillating hearts of dogs to a normal rhythm with an electric jolt. It was a revolutionary notion—electricity could be used to regulate a faulty heartbeat whenever necessary over the course of a patient's lifetime. Responsibility for the heart's most vital function, the generation of the beat, could be given over to a battery-powered gizmo. One of Harvey's heart's sovereign duties could be relegated to a benign usurper. Experimentation involving external pacemakers was taking place as early as the 1920s, although the first internal device was not implanted into a patient until Arne Larsson received one in Solna, Sweden, in 1958. That device failed after three hours; a second crashed after two days. The durable Swede would wear out twenty-six more pacemakers over the next forty-three years, surviving well into his eighties.

Although artificial pacemakers have been shrinking and becoming more efficient in the half century since Larsson's first implant, the basic concept has not changed very much. A dual-chamber pacemaker has wires that are inserted into the upper and lower chambers of the heart and screwed into the inner surface of the heart muscle. These wires are connected to a battery-powered computerized generator that is inserted beneath the skin. As with so much else in our lives, pacemakers have become increasingly programmable. Sensors can measure motion and oxygen consumption and automatically tell the heart to speed up or slow down. The device can be made to *think*. Patterns and trends can be recorded and analyzed. Dangerously fast rhythms, when detected, can be terminated with overdrive pacing, or the delivery of a shock (and it may *hurt*—science has yet to conquer this problem). And, as with devices that record palpitation, technicians can use cellular technology to monitor and interrogate a pacemaker while the patient is at home, unaware of the intervention.

Despite the effectiveness of the artificial pacemaker, there are a variety of rapid heart rhythms that cannot be treated with this device. Fast rhythms that originate above the ventricles are the most common. As scientists mapped the role of electricity in the heart, they realized that this accelerated beating is sometimes caused by an extra conduction circuit that is naturally wired into the heart. It is as if a house has a secondary electrical system that kicks on and off without warning, causing the appliances to run and lights to come on even though no one has thrown a switch. This gratuitous system can affect the heart's regular beating, doubling or even tripling it. Patients may develop shortness of breath, chest tightness, or light-headedness. In extreme cases, when the heart beats more than two hundred times per minute, the patient may even pass out. Fortunately, these episodes often terminate spontaneously. When they do not, they can be treated with drugs that block this gratuitous conduction pathway. There is also a relatively new procedure called ablation that involves frying that extra wiring with a blast of radiofrequency energy supplied by a sophisticated computer program.

The other tachyarrhythmias that occur outside the ventricle involve utter chaos. The heart's harmonic rhythm breaks down. The muscle becomes like a symphony orchestra that is constantly tuning. Early cardiologists referred to this storm of uncoordi-nated impulses as *pulsus irregularis perpetuus*; it is now known as atrial fibrillation, or afib. The concept of fibrillation destabilizes everything we are taught to think about the working heart. It is a meaningless twitch of the heart's muscular fibers, a macabre parody of its true function. In the case of afib, the atrium has been stretched or stiffened, causing the electrical wires to get jumbled. Eventually this leads to a complete breakdown in elec-trical synchrony. Rates of five hundred beats per minute may ensue. Fortunately this very rapid and irregular pattern does not fully transmit down to the pumping chambers, the ventricles. If it did, the patient would collapse and die more or less instantly. The atrioventricular node filters the signals getting through so the pulse rate is "only" in the 130 to 180 beats per minute range. Still, this can be extremely uncomfortable. And there is an addi-tional risk: fibrillation can cause small blood clots to form in the folds of the inner lining of the atria. If these dislodge, they can easily make their way to the brain and cause a stroke.

Pacemakers do not help with atrial fibrillation, and drugs are not terribly effective. Ablation is also problematic because of the sheer number of sites where this chaotic beating is occurring. But there is an alternative, one whose technical sophistication suggests the remarkable recent evolution in interventional cardi-ology. Because these errant beats tend to occur in one particular place—the back of the left atrium where the pulmonary veins connect—physicians can now lay down a ring of scar tissue around this zone and electrically contain it. Think of it as putting light-ning in a box. The extra beats cannot get to the rest of the atrium and therefore cannot trigger fibrillation.

Perhaps the greatest challenge presented by abnormal rhythms is a syndrome that has long mystified physicians and terrified just about everyone: sudden cardiac arrest. This is our deepest nightmare when we think about the heart. It defies the usual narrative of fatal illness, with its unfolding drama of symptoms, diagnosis, struggle, and acceptance. Instead, the heart simply stops pumping blood. The individual collapses without warning and dies within minutes. The end can come before we even know that it has begun.

The most common cause of sudden death is ventricular fibrillation, in which the ventricle cannot contract effectively. Ventricular fibrillation most commonly occurs when a coronary artery closes abruptly with a rapidly developing clot. The heart muscle downstream is deprived of oxygen, and the ensuing damage to the cell membranes causes an electrical discharge that can spread to the rest of the ventricle. Death occurs. Often this precedes any sensation of pain. A pacemaker is of no help—delivering an electrical charge to a heart in the midst of an electrical storm is futile. Only the delivery of a large current of energy (a "shock") or sometimes even just a thump to the chest may suffice.

Sudden death looms even for patients who survive such an attack. If the heart has been weakened and a scar from the episode remains, there is a future risk of ventricular fibrillation. A single extra beat emanating from the region of the scar can instantly spread throughout the ventricle. This can happen without warning several years after the cardiac arrest. Patients with some types of structural heart disease may also be at risk. Generally they have dilated and weakened ventricles, which may result from untreated hypertension, valve disease, chemotherapy, or pregnancy. And then there is the dreaded idiopathic cardiomyopathy—heart disease of unknown origin. Doctors suspect that it often results from exposure to a common virus, though it might just be a product of plain old bad luck.

A particularly disquieting form of heart disease that predisposes some sufferers to sudden death is known as hypertrophic cardiomyopathy, a syndrome in which the left ventricle is abnormally thickened. Occasionally it strikes down athletes while they are in the middle of a game or race. The gifted college basketball player Hank Gathers died of it in the middle of a full-court press at the age of twenty-three; top marathoner Ryan Shay, twenty-eight, was struck down during the US Olympic trials. There may be symptoms, such as a murmur, chest tightness, or shortness of breath. Or, there may be no warning. The only indication is sudden death.

Some hearts stop suddenly because they are poorly wired. There is no apparent structural disease—and yet there is a risk of arrhythmic sudden death. This faulty cabling gives us a sense of both the extraordinary intricacy of the heart's operation—and the sophistication of the threats to that function. Long QT syndrome, which often runs in families, results from faulty ion channels in the heart's cell membranes—those delicate interfaces where the heartbeat is born—that can suddenly and precipitously lead to fibrillation and cardiac arrest in young people. Wolff-Parkinson-White syndrome, first noticed in 1930 and confirmed in 1944, occurs when there is an extra circuit or pathway connecting the atria to the ventricles. The condition often shows itself with the sudden onset of a very rapid but organized rhythm that results from an extra beat traveling down the normal circuit and then back up that extra pathway. Every time this gratuitous circuit engages, the ventricles contract. The only way to cure this condition is to eliminate the accessory pathway. An electrophysiologist places a complex grid of electrode catheters at various locations in the heart. A computer then stimulates the heart to induce the arrhythmia, which discloses the location of the extra circuit, allowing doctors to zap it with a short burst of radiofrequency energy.

Bad wiring can also require a defibrillator. Most people have witnessed these devices in simulated operation, since they are used to jump-start the plots of countless television hospital dramas. The principle is simple: a short, sharp shock of electricity depolarizes the heart's chaotic activity and allows its natural pacemaker to resume the normal rhythm. The effectiveness of defibrillators is attested to by their increasing presence in a wide variety of public locations, such as restaurants and airport terminals. Consumers can even purchase them at Walmart for just over $1,250. The next big challenge is to more precisely identify patients who may benefit from prophylactic defibrillators. These can be implanted in at-risk patients who have never had a heart attack after being programmed to detect arrhythmia as it starts and correct it with an electrical jolt.

The notion of sudden heart death has been lodged in the human imagination for a long time. Literature is crowded with the corpses of characters who dropped dead from broken hearts, as we have seen with King Lear and Princess Calantha. Other forms of emotional shock have also been seen as capable of stopping the heart. In 1897, just as the modern era of cardiology was getting under way, the *New York Times* printed an article titled "D'Aumale Dies from Shock" whose subhead informs us that the deceased was "Stricken with Apoplexy on Hearing of the Duchesse d'Alencon's Death by Fire." The opening two paragraphs of the article, datelined Paris, present a picture of a world in which hearts are liable to detonate like land mines if trod upon by unwelcome news.

> The Duc d'Aumale expired at 2 o'clock this morning in his villa at Zucco, Sicily, of cardiac apoplexy. Death was caused by hearing of the death of the Duchesse d'Alençon, one of the victims of the Charity Bazaar fire.
>
> The attack which caused the death of the Duc d'Aumale only lasted a few moments and he expired without suffering. The Princess Clementine of Orleans . . . sister of the Duc d'Aumale . . . is very ill. The Princess is eighty years of age, and the shocks caused by the sudden death of her brother, following closely upon the tragic death of the Duchesse d'Alençon . . . may prove fatal.

While it is tempting to mock these fainthearted members of the European nobility from the distance of our high-tech, cardio-savvy times, it is worth noting that stress-related heart attacks are still among us. One increasingly common occurrence is known as *takotsubo* cardiomyopathy after the Japanese term for "octopus trap," the shape of which the patients' left ventricles resemble. Starting around 2000, cardiologists in Japan began to see cases in which patients, often women, suffered heart attacks after experiencing extreme emotional distress. Incidents soon began being reported in the West as well. The signs of acute coronary syndrome are all there—chest pain, elevated cardiac enzymes (whose presence in the bloodstream indicate damage to the heart), and ECG changes. There is also extensive wall motion abnormality, in which the walls of the left ventricle fail to contract normally. Though the signs and symptoms indicate a heart attack, upon examination, the patient's coronary arteries are wide open, with no plaque. Invariably the patient fully recovers, with no residual damage.

Another indication that heart disease can be stress-induced comes from recent epidemiological studies that looked at the rate of heart attacks after the earthquakes in Northridge, California, and Kobe, Japan. The results showed that the incidence of myocardial infarction in each city was much higher than on the same day the previous year. Another study had deer hunters in northern Michigan wear ECG monitors that showed changes in blood supply to the heart when they were about to shoot their first deer. (There is not yet any data on any changes in blood flow in the deer when they see the hunter raise his rifle.)

Perhaps the oddest—and most troubling—type of apparently emotion-driven heart ailment is the curious phenomenon known as sudden unexpected nocturnal death syndrome, or SUNDS. Occurring mostly in young men of Hmong ethnic origin (from parts of Asia, where it is called *dab tsuam*), it strikes the victim in his sleep, causing death in an otherwise healthy subject. Autopsies can find no evidence of disease or structural defects. The syndrome has long been explained in Hmong folklore as being caused by an evil spirit that appears in the victim's dream and scares him to death. It has also been reported among men in the Philippines, where it is known as *bangungot* and attributed to a witch called a *batibat*, who suffocates the victim by sitting on his face as he sleeps.

Not surprisingly, the syndrome caught the attention of medical researchers, who began to search for organic sources. In the 1990s the Brugada brothers, two Spanish electrophysiologists, determined that the deaths result from genetic mutations affecting the heart's conduction system. These abnormalities cause certain members of the affected populations, especially young men, to experience sudden lethal ventricular fibrillation, particularly when asleep. SUNDS became known as Brugada syndrome. Despite the name change, there is, as of yet, no treatment.

A more prosaic and much more widespread connection between the emotions and heart disease can be found in the effects of plain old daily stress. The organic changes brought about by grief, depression, and the daily grind gradually take their toll on the heart, particularly in the form of elevated blood pressure, whose devastating long-term effects on cardiac function are now well known. It is this kind of strain, rather than earthquakes, octopus traps, and nightmare-stalking sprites, that continues to wreak havoc on our hearts.

THE twentieth century's cultural representations of the heart also betrayed signs of chaos and stress. While it remained the metaphor of choice in everyday discourse, the heart also took on new figurative meanings that were often at odds with its traditional role as the embodiment of the sublime. The heart's metaphorical importance was no longer solely as a symbol of our emotional and ethical lives, or of the sacrifice of Jesus Christ. In the twentieth century, the heart also became a powerful commercial totem. The tendency that began with mass-produced

Victorian-era valentines reached a giddy culmination in 1977, when graphic artist Milton Glaser created the "I ♥ NY" rebus to promote tourism in a city stricken with fiscal crisis and crime. The entire weight of the cardiac metaphor—devotion, passion, faith—was thrown behind a campaign to get people to spend money in hotels, restaurants, and Broadway theaters. Subsequently the heart symbol became a shorthand for enthusiasm for everything from software to Yorkshire terriers. It was a stamp that validated lifestyles. People could ♥ their grandchildren or line dancing or Buddha. Although severed from its customary meanings, the heart nevertheless continued to pump out the feelings long associated with these traditions—passions that could readily be appropriated by anyone with something to sell.

There was one lifestyle above all others for which the heart symbol served as the principal logo: the pursuit of good health. Exercise, a low-fat diet, regular visits to the doctor, weight loss, reduced stress—all of these recommendations were suggested by the familiar shape. The heart symbol no longer triggered thoughts of the sublime. It reminded us that we needed to start jogging and eat fewer cheeseburgers. The clearest expression of this has been the heart-check mark that began to appear on a wide array of food packaging in 1995. The symbol consists of a heart branded with a bold, efficient check mark. It is copyrighted by the American Heart Association (AHA), which licenses it for a nominal fee to companies whose products meet the organization's criteria for saturated fat and cholesterol content. In a message to manufacturers on its Web site, the AHA asserted that "92 percent of consumers think the heart-check mark is 'important to very important' in choosing and buying foods. Shoppers are busy. They want convenience. And the heart-check mark answers their need for help in choosing healthy foods quickly and reliably." Although the AHA's use of the heart symbol is clearly a long way from that of retailers of coffee mugs and T-shirts, it nevertheless represents a cultural environment in which the heart is accorded a function beyond the dreams of Galen and Harvey: to move merchandise.

SOMETIMES a heart simply fails. It loses its ability to do the work required of it. A hundred years ago, cardiologists were talking about the heart possessing a "reserve" that would inevitably give out. Each heart had a certain number of beats allotted to it from birth. Eventually that number would be reached and the heart would just stop, like a windup clock. Now this gradual cessation is known as heart failure, and it is defined as the inability of the heart to pump an adequate amount of blood to meet the demands of the body. It is, contrary to common belief, very different from a heart attack. An attack, as the name suggests, is sudden, blinding, decisive. Failure, on the other hand, is slow, laborious, and agonizing.

For much of the twentieth century, cardiologists thought that heart failure was a breakdown of systole, the process of contraction that pumps blood into the arteries to nourish the oxygen-hungry body. In reality, a number of factors—including hypertension, coronary artery disease, valve dysfunction, viral infection, and even chemotherapy—cause the heart to enlarge, weaken, and fail. This process can easily be quantified. If one measures the volume of the heart when it is filled and when it contracts, it can be determined that, under normal circumstances, the heart ejects about 60 percent of its maximum volume. This is called the ejection fraction. In heart failure, the heart's ejection fraction is as low as 20 to 30 percent.

Physicians now understand that in as many as half of all patients with heart failure, there is no problem with the heart's ability to contract. Its systolic function is normal. The problem lies in the heart's diastolic function, its ability to fill. Conditions such as untreated hypertension can cause the ventricle's muscle to thicken and stiffen. Like its owner, it loses the ability to relax. Heart strength is normal, the ejection fraction is preserved—and yet the heart cannot fill adequately.

Whether heart failure is due to systolic or diastolic dysfunction, the clinical results, called congestion, are similar. Pressure rises within the ventricles and is transmitted back to the veins, where blood backs up like pedestrians at the bottom of a malfunctioning escalator. In the lungs, this causes water to be pushed out of the capillaries into the spaces normally filled with air. Breathlessness results, which is most evident during exertion or when the patient lies flat. In the legs, water is forced into the soft tissues and results in the swelling known as edema. Both types of heart failure—pumping and filling—can also lead to a reduction in the forward blood flow, which results in fatigue as muscles and the brain are starved of their recommended daily allowance of oxygen.

The first treatments for heart failure, notably Withering's use of foxglove, focused on trying to increase heart strength and relieve congestion. Diuretics were used to stimulate the kidneys to increase the elimination of salt and water, greatly relieving congestive symptoms. More recent treatments focus on drugs called inotropes that stimulate the heart to contract more vigorously. They can be very effective, though only in the short term. In some cases they actually increase a patient's risk of dying.

Once again, it was a change in the way of imagining the heart that allowed doctors to tackle the problem most effectively. Think of a horse-drawn cart loaded with heavy barrels being pulled up a steep hill. When the driver realizes the horse cannot make it, he has two choices. One is to whip the horse. The other is to throw some of the barrels off the cart. (Actually, there is a third choice: getting a new horse. Transplantation will be discussed later in this chapter.) Until the latter part of the twentieth century, treatment for heart failure focused on finding ways to whip the horse harder. Recently, however, doctors have begun to take a more effective approach. They have begun to unload the cart. By asking the heart to do less rather than forcing it to do more, they have begun to solve one of the most intractable problems in cardiology.

This therapy focuses on an unlikely culprit—the kidneys. The heart and kidneys are more intimately connected than is commonly supposed. The kidneys not only eliminate waste products, they also are critical in managing blood pressure and volume. When blood flow is reduced, they release a hormone that causes blood vessels to constrict and salt and water to be retained. Early on in heart failure, these compensatory changes help maintain blood flow. Ultimately, however, they overshoot. The narrowing of the body's blood vessels makes it harder for the heart to eject blood, and salt and water retention exacerbates congestion. To make matters worse, the adrenal glands get in on the act and secrete adrenaline to try to whip the heart to greater effort.

Reversing these compensatory changes is the key to treating heart failure. Drugs called ACE (angiotensin-converting enzyme) inhibitors dilate the arteries, allowing the heart to pump blood more easily. Like many pharmaceutical breakthroughs, this class of drugs is based on a naturally occurring compound—in this case a chemical found in the venom of a Brazilian viper. These medications clearly improve symptoms and survival in patients with heart failure, though often they are not enough.

Adrenaline must also be dealt with. As anyone who has ever been terrified can attest, adrenaline is a potent stimulant that increases heart rate, blood pressure, and the contractile strength of the ventricle. This might seem like a good thing in a patient with heart failure. The problem is that adrenaline also causes the arteries and veins to constrict. It is like increasing the flow of water while at the same time narrowing the gauge of the pipe. Something has to give.

The solution to this problem of adrenaline excess came about in one of those daring leaps of imagination that seem to characterize the whole history of cardiology. It involved beta-blockers, drugs that can inhibit the effects of adrenaline. Because limiting adrenaline lessens the heart's pumping strength, these drugs would seem to be the last thing a doctor would want to give a patient suffering from heart failure. But reversing the toxic effects of adrenaline was of greater urgency, so doctors gambled by giving beta-blockers to patients suffering from heart failure. And it worked. If the therapy starts with low doses that are very gradually increased over time, many patients have a remarkable recovery in heart strength and do, in fact, live longer.

Surgery on the failing heart to improve its contractile strength has not proved very effective. Attempts to excise weak portions of the ventricle and reshape it, or to wrap skeletal muscle around the heart, have not worked. Transplantation is by far the most effective surgical remedy. Contrary to popular belief, from a technical standpoint transplantation is one of the simplest operations a cardiac surgeon performs. The anatomical and physiological challenges of the procedure were solved well before it was first accomplished. As early as 1905, the dexterous Alexis Carrel transplanted the heart of a puppy into an adult dog. It beat for nearly two hours. The problem, it turned out, was not one of surgical skill or scientific knowledge; it was rejection. When confronted with a new heart, the body will naturally deploy its formidable immune system against it. Transplantation appeared to be the ultimate Pyrrhic victory. In defeating the invader, the body killed itself. Although South African cardiothoracic surgeon Christiaan Barnard's breakthrough 1967 transplant operation is the stuff of legend, less well known is the fact that his onetime teacher, Stanford University's Norman Shumway, had been capable of performing the procedure for several years but was holding off until effective immunosuppressive therapy was in place. His caution was well advised. Barnard's trailblazing coup brought about a gold rush that turned up mostly pyrite. Although there were almost a hundred heart transplants performed worldwide in 1968, three years later there were fewer than ten. The rejection rate was too high. Bodies were simply not accepting the transplanted organs.

The solution to the rejection problem occurred almost by accident. The biggest breakthrough in cardiac transplantation came not in an operating theater, but in a remote patch of Norwegian forest where a Swiss microbiologist collected a batch of soil while on vacation, hoping to unearth new antibiotics. Upon returning to work, he and his colleagues eventually isolated a fungus that yielded the drug cyclosporine. Although it was a bust as an antibiotic, the substance turned out to be remarkably effective at warding off T cells, the commandos in the immune system's assault on the transplanted organ. By the 1980s this wonder drug had led to a resurgence in cardiac transplantation. By the first decade of the twenty-first century, almost four thousand such procedures were being performed worldwide each year, with an average survival time of fifteen years. The biggest obstacle to even greater numbers is no longer medical know-how, but a lack of donors. Healthy, genetically matched hearts that are still beating in brain-dead patients are relatively rare. And they do not keep on ice like kidneys and corneas.

Whatever its ethics or outcome, the first heart transplant rivaled the moon walk that occurred two years later in its effect on the human imagination. A barrier that had once seemed absolute had been breached. But it was more than just a scientific breakthrough. Barnard's surgical feat undermined the way that Westerners had been thinking about the heart for nearly three millennia. The scientific line connecting Harvey's lectures with that Cape Town hospital may have been relatively straightforward, but the metaphorical rupture between these two milestones was profound. How could the heart be the locus of our deepest selves if it could be completely replaced? How could it be sovereign if it could be so comprehensively overthrown? Is a heart recipient the same person when she wakes up as she was before she went under the knife? Does she love her husband in the same way? Are her deepest secrets still her own?

Our faith that each of us possesses a unique, individual heart came under an even more profound challenge in 1984, when a team of surgeons in Loma Linda, California, transplanted the heart of a baboon into a twelve-day-old infant girl known as Baby Fae. She survived for three weeks. A peer reviewer later accused the surgeons of indulging in "wishful thinking" for believing that a human body would accept an organ from another mammal. Xenotransplantation, as cross-species organ grafting is known, remains an approach that most doctors avoid like the plague—literally. In addition to rejection issues, there is also the possibility of importing animal diseases such as those caused by retroviruses into the human species. Conceivably, both of these problems can be resolved. The ethical and religious issues, however, will remain, as will the crisis in the human imagination that will be triggered by a successful xenotransplantation.

IN the post-Barnard era, old ways of thinking about the heart came under serious attack. The imagination fought back, stubbornly holding on to old paradigms and beliefs. Myths grew up in which heart transplant recipients absorbed the thoughts, memories, and emotions of the unknown donors. Urban legends began to be told of patients dreaming with almost photographic accuracy about the accidents that had taken their donors' lives, even though they had no way of knowing these details. Recipients allegedly underwent personality changes that could not be explained by the usual postoperative trauma. These narratives seek to merge some of the oldest thinking about the heart with some of the very newest. Whatever the particulars of each case, the story is the same: something happens to our essential selves when we take on another person's heart, a transformation that does not occur with transplantation of other organs.

The 2003 film *21 Grams* vividly illustrates this tension. Made by the gifted Mexican team of director Alejandro González Iñárritu and writer Guillermo Arriaga, it tells the story of Paul (played by Sean Penn), a dying mathematics professor who receives the heart of a man who is accidentally run down by Jack (Benicio Del Toro), an ex-con who has become a born-again Christian. Paul, who initially makes a strong recovery, soon becomes obsessed with the idea of connecting with the victim's grief-stricken wife, Cristina (Naomi Watts). He even goes so far as to hire a private investigator to track her down. Although Cristina—whose two young daughters also died in the accident—is initially repulsed by the idea of dealing in any way with this stranger who carries her husband's heart, she soon enters into an intense sexual relationship with him. Their affair culminates in her enlisting Paul to kill Jack, who was released from prison after serving a short sentence.

With Paul, the filmmakers provide a striking commentary on our conflicted view of the heart. As a man of science, he is initially at ease with having his heart treated as a lump of dying tissue. After it is replaced, he asks his doctor if he can see it, though he displays no sense of reverence when he sets eyes upon it. He even jokingly calls it "the culprit."

This cavalier attitude soon changes. Paul's new heart begins to drive his actions in ways that confuse and frighten him. He starts to take extraordinary measures to adopt the life of the man whose heart he carries. He stalks and then seduces Cristina by telling her that she does not have to be afraid of him. "I've got a good heart," he assures her. He even agrees to join her in premeditated murder after she tells him that it is his duty to avenge the man whose death made his life possible. "You owe it to [my husband]," she says. "You have his heart." When Paul cannot bring himself to kill Jack, he instead shoots himself in his new heart, which by this time is in the process of being rejected. It is as if he believes that his cowardice makes him unworthy of the other man's heart.

What is most intriguing about the film is that its makers never identify the source of Paul's obsession with taking on the life of his donor. Is it merely a manifestation of the guilt he feels after learning the unbearably painful circumstances that brought him the new heart? Or does the organ's implantation actually cause him to fall in love with Cristina? Is he suffering from an elaborate form of postoperative stress? Or has this alien heart given him a new emotional identity in addition to a new biological one?

Despite the headlong march of science, these questions clearly remain alive in the era of the transplant. After all, it is hard to imagine the film being made with the same story line, but with Paul receiving a new liver. Although the filmmakers leave the viewer room to believe that Paul's motivations are simply a matter of his own delusions, the alternative explanation can also be defended. The heart remains a potent metaphor for devotion, faith, and intense feelings. Even the woebegone Jack participates in this symbolism—he bears a vivid sacred-heart tattoo on his neck, which causes him to get fired from his job, an event that ultimately puts him at the scene of the accident. Even in the form of a common tattoo, the heart's metaphorical resiliency remains in place.

Another radical challenge to how we think about the heart comes from a therapy that threatens to overtake transplantation as the preferred means of treating heart failure: the insertion of a synthetic device that will perform the heart's duty. Although the popular imagination usually thinks of this technology in terms of an artificial heart that completely replaces the biological organ, the current trend is toward left ventricular assist devices (LVADs), which are positioned near a failing heart to take up the job of pumping blood while leaving the muscle, which may still be functioning somewhat, in place.

The first artificial heart was, technically speaking, Gibbon's heart-lung machine. The millions of people who have undergone heart surgery since its creation have temporarily possessed artificial hearts, albeit externally. The first internal artificial heart was implanted in 1969 in Houston by a team led by Denton Cooley, though it was only a "bridge to transplantation" and was removed after three days. In December 1982, fifteen years after Barnard's legendary transplant, a team in Utah installed the Jarvik-7 artificial heart into a retired dentist named Barney Clark. It was the first artificial heart intended to be a permanent feature of the recipient's anatomy. Clark died after 112 days. Since then, there has been no significant progress in implanting fully artificial hearts. Blood clots and infection remain daunting challenges, though a French team of transplant specialists and venture capitalists is reportedly at work on an artificial heart that will deal with rejection issues with a biosynthetic pseudoskin made of chemically treated animal tissues.

Much more progress has been made with LVADs. They can be used as bridges to transplantation or to replace heart function for the remainder of a patient's life. The natural heart need not be removed. A recent case in England suggests another potential use for these devices. Doctors had implanted a heart in a toddler whose own organ was too weak to keep her alive, though they left the original heart in place. Ten years later, they were able to remove the donor heart when the patient's own heart had recovered (having had what her surgeon called an extended time-out). This "piggyback" transplant opens up a whole new frontier in the treatment of cardiac failure, in which deteriorating hearts are given periods to rest and recover while either transplants or LVADs do the heavy lifting. The old image of the redeemed heart is taking on an entirely new meaning. This tantalizing development provides a view of the future of heart care, in which cutting-edge technology is used in conjunction with the heart's remarkable natural capacities to treat diseases that neither force could heal on its own.

CHAPTER 6

Future Heart

Washington State, 2021

Kirk

He was born and raised in a small farming town in the eastern foothills of the Cascades. Washington was a beautiful place to grow up, but there is not much in the way of work in his town. Most of his friends take jobs farming wheat and hops. The only other alternative is working in the slaughterhouse in nearby Ellensburg. Not eager for either of these options, Kirk spends two years studying English at Central Washington University in Ellensburg, thinking that he might become a teacher. But he drops out when he can no longer afford the tuition. Although he was on the varsity basketball team in high school, he started smoking when he was sixteen and was soon up to a pack a day. By the time he leaves college, he can see the foolishness of it. But he's already hooked.

Besides, he's got too much on his mind to try to stop smoking. After leaving school, he works as a landscaper in the summer and drives a cab in the winter. Then he marries Laura, his high school sweetheart. She learns to cut hair in the local beauty salon that she will eventually own with a partner. They make enough to get by. He drives a pickup with 150,000 miles on it. They are always talking about going down to Baja for a vacation, but there never seems to be enough money.

Despite the cigarettes and the long hours, Kirk is healthy. He can ride his bicycle a hundred miles on any given day. He almost never misses work. He never sees any reason to visit the doctor and certainly is not going to shell out a small fortune for health insurance. If things get really hairy, there is always the ER. That's what just about everybody he knows does. He doesn't spend much time worrying about this, however. He has Laura and he has his health. That's all he really needs.

Meanwhile, he can only watch helplessly as his sleepy town changes. When he was a boy, the prosperity of the 1990s started to bring in a new class of residents. This really picks up when Kirk is in his thirties. Yuppies from Seattle or California cash in their Microsoft or Intel shares and buy up the land. The wheat and hops are plowed under, replaced by grapes for making chardonnay and pinot gris. The local town becomes packed with cafés and gourmet restaurants and high-end boutiques. The cost of living skyrockets. You have to drive out to Applebee's at the new strip mall to be able to afford to eat out.

Kirk never really benefits from any of this "progress," except that it allows him and Laura to work longer hours so they can make ends meet. There is always plenty of landscaping work at the big estates the newcomers from Seattle and California build in the foothills. Soon Kirk has time for only the drive-through meals that are beginning to form the bulk of his diet. He rarely rides the bike anymore, and when he does, he puffs and pants after just a few miles. His waistline starts to show it.

And somehow along the way, he manages to upgrade his smoking habit to two packs a day.

Raj

He retires when he is thirty-four, which is twenty years younger than his father was when he stopped working, though in the old man's case, retirement came in the form of dropping dead of his second heart attack. Not that anyone was too surprised by that. His father had been working fourteen-hour days since arriving from Calcutta as a teenager, first at his uncle's convenience store and then as an accountant after graduating from night school. At least he got to see Raj get into Stanford. If only he could have held on a few more years—then Raj could have gotten him out of the rat race. God knows there is enough money now that he has cashed in his stock options. More than enough.

After Raj leaves his job, he moves out of Seattle with his wife, Monica Waterman Singh, and their daughter, Saffron. The 299 acres they buy have prime soil for growing pinot gris grapes. The place is paradise for them. Raj works from dawn until dusk to bring in the first vintage, but after that he turns the work over to experts. He starts doing some consulting work for his old company. Strange as it seems, he misses the drama of writing code—the sixteen-hour days and impossible deadlines. Monica, who was in marketing before the move, soon grows restless as well. Although she's initially a whirlwind about managing the army of interior designers and landscapers they hire to build their dream house, before long she is talking about starting up an educational-software firm to market some of the ideas she gets from homeschooling Saffron.

Raj is supposed to be living easy now that he has cashed in his chips, but he sometimes wakes up in the middle of the night, his heart racing with a panic whose source he cannot put his finger on. Because they have not found a new doctor they like—this place is beautiful, but it's also *remote*—Monica makes him call their internist back in Redmond. He has Raj send in his ECG via the iGram app on his phone. All you have to do is hold it against your chest for a minute, then press Send. The doctor reads the results and tells him he is fine, though he should probably have a scan the next time he's in Seattle, just to be safe. And he should start exercising more, maybe take up a hobby. Or charity work. To help Raj chill out, the doctor sends him a prescription for the latest issue of Ambien.

Kirk

Everything changes when his older sister, Margaret, dies. She was the one who raised him. Her operation is supposed to be a routine hysterectomy—one night in County Hospital. But the day after she gets home, she suddenly cannot breathe. She calls Kirk on his cell phone, but he can scarcely make out her words. He races to her house, forgetting that he has a fare in the back of his cab. By the time he arrives, she is already dead. The autopsy says it was a pulmonary embolus, a massive blood clot in a lung, most likely from a pelvic vein. A one-in-a-thousand chance, they say, although rumor has it that the HMO doctors at County are overworked and undermotivated, making Kirk wonder if they were negligent in her care. But there is no way of proving any of this.

This starts Kirk's downward slide. It is fast and it is steep. Alcohol helps a little, but the hangovers are miserable. OxyContin proves easy to get but makes him miserably nauseous. Cocaine, on the other hand—now, *that* is just what the doctor ordered. The energy is exhilarating, and a sixteen-hour workday is a piece of cake. He uses the coke to power him through a big landscaping job for some nice, young, rich people in the foothills, an Indian guy and his nervous blonde wife. Cute little girl. What is her name? Something unusual. And then there is the clarity of thought the cocaine brings. He starts writing again—poetry and short stories, even an illustrated Christmas book for his cherished nieces. So what if he crashes from time to time? The depression is only temporary. A little boost always rights the ship.

But then the headaches start. A sledgehammer pounding his temples, and blurry vision. He keeps doing lines, though it is getting harder and harder to work. Once, when he stops at a drugstore for some Advil, he tucks his arm in the blood pressure machine on a whim: 220/130. No way. He does it again: 222/128. These damn things never work, he thinks. Though something deep in his mind tells him he should see a doctor. The problem is, he doesn't have a doctor. And he's seen what his sister's surgery did to her family's finances. The HMO nickel-and-dimed them to death. Deductibles, co-pays. What do they call them? "Out-of-pocket expenses." Yeah, that's right where those guys have their fat, greedy hands—in your pocket. Getting involved with doctors and insurance companies is the last thing he needs, what with all the money he is blowing. In more ways than one.

But snorting cocaine is no longer getting him through his shift. So he starts smoking it. The highs are amazing. But then there is the chest pain. Laura thinks it's because of all the crap he eats, but this isn't heartburn. After ten years of nonstop fast food and all-night shifts, he knows heartburn. This feels more like something inside of him is tearing, ripping straight through to his shoulder blades. He knows this is serious, but what can he do? A trip to the ER is, what, two grand minimum? So he puts up with it for a couple days, hoping it will just go away.

And then, late one night, as he's sitting in his empty cab, the pain becomes unbearable. He never knew pain could be this bad. He heads straight home. He is just coming through the front door when he collapses.

Raj

The sleeping pills help. Finding a hobby proves more difficult. He signs on to teach basic computing at the local high school, but it turns out to be a horror show. He'd thought things were bad when he was in public high school back in Oakland, but that was nowhere near as grim as this. They still use desktops! It's like the Stone Age out here.

And then it happens. A month after his initial call to his Seattle doctor. Right out of the blue. He wakes up in the middle of a cold winter night with the most incredible pain he has ever felt. It is as if someone is ripping the bone and muscle right from his chest. He wakes his wife; she calls their internist. Monica has to hold the iPhone to Raj's chest.

"Call an ambulance," the doctor orders. "Do you have a HeartStart in the house?"

"It's in the Lexus."

"Good."

But Monica does not need to use the home defibrillator. Nor does she call 911. They would probably just take him to County

Hospital, and she's heard plenty of stories about that place. Instead she calls the 24-7 number of their customer-care specialist at MedGroup Platinum. She's never been happier that Raj negotiated carrying his corporate health insurance over as part of his severance deal. A human being answers on the first ring; he knows just what to do. The private ambulance arrives in fifteen minutes. It races them to the MedGroup remote tertiary-care facility that opened in the valley two years earlier. Although the small, unmarked building looks like just another office complex from the outside, it turns out to be a fully equipped, fully staffed medical facility—so high-tech it could be a control station for a Mars mission. And there is no wait.

They immediately perform an MRI. The news is grave: aortic dissection. The inner lining of the main artery in Raj's chest has torn away from the outer shell. According to the radiologist, a rupture is imminent. And the nearest vascular surgeon is in Seattle, a two-hour drive over a mountain pass. To complicate things further, a fast-moving winter storm is blowing in from the Pacific. Those roads are tricky at the best of times—if the storm hits, they will become downright perilous. A MedGroup chopper could get him there in twenty-five minutes, though it could be grounded the moment it arrives at the local airport because of the storm.

There is, however, an alternative.

Kirk

The ambulance takes forever to arrive. One of the paramedics turns out to be an old high school friend of Kirk's, Wes. They played basketball together. He rushes Kirk to County as quickly as possible. By the time they get him into the ER, he is slipping in and out of consciousness and barely has a pulse. To make matters worse, there is a backup in radiology. When they finally do a scan, the news is bad. Something called aortic dissection.

It turns out that the nearest surgeons who can handle this are in Seattle. Normally, the staff at County would see if they could get him there on the state's medical chopper, though you never knew where the hell that thing was. Could be hours before it is available. Half the time, it's broken down. Besides, in this weather it will probably be grounded, which means driving. The problem is, it will be tough to find someone who's willing to go over the mountains with that storm rolling in.

"I'll do it," Wes says wearily. "What the hell. Go Bulldogs."

And so off they go. Although they stabilized Kirk in the ER, Laura can see he is in bad shape. If the aorta ruptures, he'll bleed to death in seconds. She wills Wes to drive faster, but she also wants him to be careful not to hit any black ice. Although it is snowing pretty hard, they make good time. Until they reach Snoqualmie Pass.

"Damn," Wes mutters beneath his breath.

Laura looks out the front windshield. All she can see is white.

Raj

They wheel him straight into the virtual OR. He is prepped for surgery, a relatively simple process that consists of inserting a catheter into the artery in his groin. The satellite link with Seattle is established and the robotic arm is positioned as the anesthesiologist puts Raj under. The surgeon in Seattle, wearing stereoscopic headgear and manipulating a joystick that was developed by programmers from Nintendo and Intuitive Surgical, uses the catheter to place an endograft in the patient's torn aorta. Although the surgeon is a hundred miles away, he is able to thread the collapsed metallic stent through Raj's arteries to the site of the bulging aneurysm without difficulty. Once in place, the endograft is expanded, causing it to grip the arterial wall at both ends of the aneurysm, bypassing the bulge from the inside. Next, the surgeon repairs the leaking aortic valve.

A cardiologist then takes over from his office in Eugene, Oregon, using the same technology to neatly slide a bioabsorbable stent into the left main coronary artery that has been partially occluded by the flap loosened by the aorta's dissection. It opens the vessel efficiently; a few years down the road, it will have merged with the artery to form a new, stronger vessel. And that is it. The bleeding has been stopped. The patient might be in the middle of nowhere, but he is out of the woods. When Raj is allowed to see his wife and daughter, he thinks they have never looked so beautiful.

His treatment is not over, however. As they work on Raj, the doctors can see that the front wall of his left ventricle is shot. This is not the sort of problem it would have been to repair twenty or even ten years earlier, however. They explain to Monica that they are going to put her husband on an assist device for the next couple of weeks. This is a small external pump that will take over the workload of the heart. He can take it home with him. Before he leaves the virtual OR, a technician harvests skin cells from Raj's leg to ship to a biotech company in Sunnyvale, California. The technicians there will transform the cells into myocardium with a little genetic hocus-pocus and then return them so they can be injected into the damaged area. In short order the scar tissue will be replaced by muscle that would rival that of a sixteen-year-old.

Raj will be just fine. The only miscue comes a few weeks later, when a bill arrives from the care facility demanding a $750 surgical co-pay. Monica calls their customer-service representative and tells him that it was her understanding that there would be no out-of-pocket expenses with the Platinum plan. The rep immediately agrees to remove the mistakenly applied charge. He could not be nicer about it.

Kirk

The two-hour drive to Seattle takes almost six. By any reasonable standard, Wes should turn back. But he plows on. Three times he has to stop to dig them out. The last time, a passing trucker helps pull them free with a chain. By the time they get to the hospital in Seattle, Kirk barely has a pulse. The surgeon has to finish operating on someone remotely by using some sort of robot, so there's a short wait, but then they are good to go. Kirk is in there for a long time. It's touch and go, but the surgeon does manage to repair the aorta in the end. The news is not good, however. So much damage has been done that Kirk's chances of long-term survival are slim. If he had arrived at the hospital sooner, they could have used some new procedures—and the surgeon has heard a rumor that the HMO might even approve some of them. Laura doesn't bother to tell him that they have no HMO. People will find that out soon enough.

Kirk dies after one more operation and six weeks in intensive care. The bill comes to $2.2 million. In a way, Laura is glad it is so big. If it had been a smaller number, she might have felt obligated to pay them something. She sells the salon to her partner before the collectors can come after it. Of course, she is able to stay on as an employee. The wages are low, but there are always tips. She uses some of the money from the sale to pay Wes for his heroic mountain run. The rest she puts toward the funeral. Luckily, with winters having become so short, they don't have to wait until spring to bury Kirk next to his beloved sister.

HIPPOCRATES just might be proved right in the end. Perhaps when we think about the heart in days to come, we will see it as being largely immune to disease. There appears to be no limit to advances in cardiology. It is now entirely possible to imagine a time when a heart can be preprogrammed in the womb to prevent all manner of ailments and defects, when most of its parts will be replaceable by mass-produced components whose life expectancy can be measured in decades, even centuries. One can even picture a time when the fatal heart attack will be as rare as the lightning strike to which it was once likened.

These are no longer fanciful notions. The fifty years following the creation of the heart-lung machine saw astonishing progress in our ability to heal the heart, and the advances set to occur in the next half century are almost impossible to fathom. Bioengineered drugs, increasingly precise imaging, xenotransplantation, gene therapy, robotic surgery, regenerative tissues, biologic pacemakers, ever-more-sophisticated assist devices—it is tempting to think that there will come a time when physicians will be able to manage serious heart ailments so successfully that they will cease to be major causes of death and disability. The future heart will no longer be a source of awe or wonder or dread, but just another organ. A kidney with a sense of rhythm. Even its most mysterious feature, the pulse, will be under complete human control, bolstered by an internal mechanical pacemaker that can be monitored by a cell phone and adjusted, if need be, by a technician on the other side of the world.

Two forces currently at work could slow the arrival of this new era, however. The first is human nature. Despite being bombarded with warnings and exhortations, statistics and targets, a large number of us stubbornly continue to harm our hearts by engaging in an array of avoidable behaviors. We consume fat-rich diets and smoke tobacco; we refuse to exercise and do nothing to alleviate the stress piled on us by an increasingly hectic world. A substantial part of the population is simply not getting the message that we can do much to control our own destinies when it comes to heart disease. Although medical science is becoming more adept at treating the fallout from these lifestyles, avoiding them in the first place will always be preferable to a trip to the doctor after the damage has been done.

The second complicating factor is cost. Gleaming new technologies and cunning genetic therapies are only so much science fiction to a patient without the means to afford them. More needs to be done to make sophisticated cardiac care available to the disadvantaged, or else heady breakthroughs in heart treatment threaten to remain beyond the reach of many while a fortunate few receive cutting-edge treatment for coronary artery disease, hypertension, and other syndromes. While Congress's passage of health care reform legislation in 2010 suggests that there is at least some political will to tackle this problem, the gap between what science can do and what society can afford threatens to grow even more pronounced in the foreseeable future.

Whatever the social and economic circumstances, scientific progress will continue. One area where this will be seen clearly—literally—is in increasingly accurate images of the heart. CT, MRI, and ultrasound technologies are constantly being refined, creating cardiac portraits with increasingly exquisite anatomical detail. Coronary angiography, with its ultrasonic transducer-tipped catheter revealing every nuanced facet of the inner walls of blood vessels, will continue in the near term to be the technology of choice for visualizing obstructions. And yet, for all its utility, this technique still requires placing a catheter into an artery, which entails a certain amount of risk. Eventually, noninvasive techniques will provide equally high-quality imagery. The heart and the vascular system will become utterly transparent. One intriguing new method is the PET scan, in which a positron-emitting isotope is introduced into the bloodstream and then taken up by heart-muscle cells. The heart is illuminated. A specialized camera detects the positron, creating a sort of X-ray from the inside out. Sophisticated software then reconstructs a portrait of the heart that can yield more information about its size, strength, motion, blood flow, viability, and scarring than ever before.

Despite the scrupulously accurate images that such techniques can yield, they still can visualize plaque only after it is very advanced. Detecting plaque at earlier stages of development is critical. Physicians are getting a better view of this process with a technique called electron-beam computed tomography, which enables them to see coronary calcification at increasingly early phases and allows them to tell patients sooner to get their acts together. Another method, a carotid intima-media thickness test, uses ultrasound to scan the carotid arteries in the neck. If a vessel wall has started to thicken, that is a reliable indication that plaque is being deposited elsewhere.

The next generation of imaging will be not just diagnostic by detecting plaque, but also prognostic. Physicians want to be able to see the heart's future. The holy grail of this type of cardiac screening is a test that detects the specific type of vulnerable plaque that is liable to rupture at any moment. Like the animated radar maps used to illustrate coming storms, these devices will show us not pictures of the relatively clear sky we see through the window, but rather the dangerous fronts that are bearing down on us.

Drugs are another area in which major breakthroughs will continue to be made. The current array of statins and beta-blockers will give way to a whole new pharmacopoeia. This is not to say that some of the old standards won't remain in use. Although it is tempting to think that medicine has come a long way since the days of foxglove and leeches, doctors continue to prescribe digitalis and bivalirudin on a regular basis. Aspirin is also going to be tough to beat as a standard anticoagulant. In fact, the real pharmaceutical revolution in treating heart disease will not involve finding new drugs to replace the old ones. It will involve *making* new medicines. The days when pharmacologists milk snakes for their venom and slog through peat bogs in search of the next wonder drug are winding down. Science is moving from discovery to engineering, from chemistry to biology.

In cardiology, this will be most readily apparent in the way drugs are delivered to the heart. Engineered drugs will be able to penetrate to their targets much more efficiently. One promising method involves a kind of Trojan horse strategy in which drugs are placed inside phospholipid envelopes, which are basically fat droplets that can sneak medicines through to the same places that harmful cholesterol now reaches. Nanotechnology will also be used to parcel out drugs into extremely small particles so they can more easily percolate to their intended destinations. To get a sense of exactly how small some of these new pills will be, keep in mind that a nanometer—the basic unit of measurement used in manufacturing them—is one-billionth of a meter.

But designing drugs from scratch can be costly, laborious—and humbling. Researchers are constantly recognizing that they simply are not smart enough to create the medications that physicians need. Their equipment and software are no match for the sophistication of the diseases they are fighting. Fortunately, there is another laboratory in which they can work, one that is equipped with a mind-boggling array of equipment and technologies: the body. With the expansion of a field known as biologic therapy, scientists are increasingly relying on living organisms—animal and human—to help them create effective medicines. The body is becoming a factory for manufacturing the raw materials needed to make disease-fighting tools.

An example of this growing discipline is personalized medicine. Although its primary application is currently in the fight against cancer, it will also have important uses in cardiology. In personalized medicine, each patient's DNA is decoded to inform doctors of what the most beneficial therapy for that particular body will be. Instead of simply taking some standardized measurements and an oral history, the doctor will read the patient's genomic printout in the same way a mechanic would consult the specs of a sophisticated machine in need of repair. This will allow the physician to abandon the one-size-fits-all approach to therapy for a minutely calibrated approach customized to the patient. The book of the heart will become a repair manual that runs to thousands, if not millions, of pages.

Perhaps the most important immediate use of this personalized cardiology will be pharmacological. Warfarin, or Coumadin, for example, is a blood thinner whose benefits have long been recognized. (Its benefits for *human beings*, that is—warfarin is also the active ingredient in rat poison.) At the right blood level, it can prevent clots that could dislodge and cause a stroke. But the wrong dosage can be dangerous, or even fatal. Because of the drug's unpredictability, cardiologists have never been able to know in advance the precise amount to administer to a patient. Ballpark guesses are the best they can do, and these occasionally strike out badly. Constant monitoring is mandatory, with clotting tests often needed every few days. Even so, thousands of patients taking warfarin wind up in the hospital each year to receive corrective treatment. With a genomic profile, however, cardiologists will be able to know more precisely the amount of warfarin required to prevent clotting without causing bleeding. This strategy will eventually expand to include a wide variety of other medicines and measures, removing much of the speculation from an undertaking in which one wrong guess can prove catastrophic.

Antibody therapy is another field that has a promising future in treating heart disease. Once perfected, it will allow scientists to create magic bullets that will attack specific targets in the human body, not only making medications more effective, but also diminishing the chances of unwanted side effects. The bullet in this case is an antibody, a protein created by white blood cells that is used by the body's immune system to neutralize foreign objects such as viruses and bacteria. Evolution has turned antibodies into weapons with deadly aim.

To customize this weapon to fight heart disease, scientists have enlisted the help of a hairless, laboratory-bred creature known as the nude mouse. Born with a defective thymus, this woebegone rodent possesses a radically suppressed immune system, allowing it to be used as a vehicle into which scientists can graft a wide spectrum of tissues without worrying about their being rejected. These tissues can be taken from other mice as well as from human beings. Scientists can then use this implanted tissue as a source for cell lines that will allow them to manufacture dedicated antibodies to target specific diseases.

This therapy is already showing promise in treating diseases ranging from breast cancer to rheumatoid arthritis. One of its main uses for the heart is to create magic bullets that will target platelets that form clots in the arteries, breaking them up before they can turn deadly. Another use will be to attack the inflammatory proteins that can wreak havoc in the heart muscle after a stent has been inserted. Because antibodies from mouse cell lines are, for obvious reasons, liable to cause allergic reactions when introduced into the human body, scientists have come up with an ingenious way around this obstacle. Once they have manufactured the mouse antibody, they chop off one end and splice on a fragment of a human antibody. The resulting hybrid organism is called a chimeric monoclonal antibody, named for the Chimera, the mythological beast made from a lion's head, a goat's body, and a serpent's tail. This "humanized" bullet is then shot into the patient's body, which recognizes it as belonging to its own species and lets it proceed unmolested to the targeted cell to do its work.

Gene therapy will also become crucial to the future of heart care. It has less to do with treating heart diseases than with keeping them from happening in the first place. The idea, quite simply, is to alter the genomic structure to build a better heart, one that can fend for itself. For example, physicians are now eagerly looking forward to the day when they will be able to insert the gene that codes for the LDL receptor in all of their patients. When the gene is faulty, high blood cholesterol can develop even in people who exercise and watch their diet. Altering it will, in effect, make the body able to police itself by sweeping bad cholesterol from the blood. It will also lessen the current heavy reliance on statin drugs, whose complications range from muscle aches to liver damage requiring transplantation.

Another important genetic breakthrough came from the small northern Italian town of Limone sul Garda, where researchers discovered a family that is more or less immune to coronary artery disease. What is even more unusual about them is that they have lower-than-average levels of HDL (good) cholesterol, which is the opposite of what researchers would expect. It turns out that what they *do* possess is a type of HDL whose particles are much larger than normal. Scientists have been able to copy this unique gene—which they call HDL Milano—and are now working on a way to radically extend this lucky family's boon (and an awful lot of lives) by infusing it into the rest of us.

Angiogenesis, the process that grows new blood vessels, is an area that is also positioned to undergo tremendous expansion. Future therapies will have their origins in cancer treatment, where scientists are looking for ways to cut off the blood supply to tumors. Heart researchers, however, want to do just the opposite—and they believe that they will soon be able to stimulate the heart to grow new arteries. They have isolated the gene for growth factors, proteins that kindle the manufacture of new blood vessels. Unfortunately, the ever-present problem of the patient's immune system rears its formidable head when these suspicious molecules are introduced. Once again, scientists are solving the problem with packaging, this time placing the DNA for growth factors inside weakened adenoviruses, the bugs that cause the common cold. They can then inject millions of these attenuated viruses into the bloodstream within a coronary artery, where a number of these hybrids will then penetrate the artery's walls and infuse cells with the new DNA, enabling them to create new blood vessels.

Another extremely fertile area of research is stem cells, the unspecialized cells that can differentiate into other types of cells to regenerate damaged tissue. Embryos are a rich source of these cells—as well as a rich source of controversy. Researchers plan to bypass these ethical and political difficulties by harvesting a patient's own skin, skeletal muscle, and bone marrow as sources of stem cells. In a laboratory, the cells can then be directed to differentiate into cardiomyocytes—heart-muscle cells—and trained to beat together in a coordinated fashion. Once they are all marching to the same beat, the cells can then be injected into a damaged area of the heart. Scar tissue is replaced with new muscle. The heart muscle, once believed to be incapable of regeneration, is reborn from the patient's own cells. An even more ambitious program will have cardiologists directing stem cells to differentiate into conduction-system cells, thereby creating a biologic pacemaker. The patient's heartbeat will be conjured in a petri dish.

We can also expect bold developments in tissue engineering, which involves creating tubes from elastic, biocompatible materials and then lining them with the patient's own endothelium—the layer of flat cells coating the insides of blood vessels—that has been grown in the lab. If successful, this approach would provide a much better bypass graft than the vein sections that are currently removed from the legs. As noted in Chapter 5, tissue engineering is also being applied in artificial heart research. A patient's own cells could be used to create a skin of sorts that will insulate assist devices from attack by the body's immune system.

Stents will remain a central platform of interventional cardiology for the foreseeable future. They are simply too effective at opening severely blocked arteries to abandon anytime soon. But there are limitations. When a cardiologist places an eighteen-millimeter-long stent in an occluded left anterior descending coronary artery, he may be saving a life in the near term, but he is also treating less than 5 percent of the owner of that vascular system's vessels. Nothing is being done to halt or reverse the underlying process of plaque deposition, which is surely also occurring elsewhere. Another limitation is that a stent, for all its technological virtuosity, is still little more than a simple metal tube. It is essentially a piece of shrapnel. It often does not slide down a tortuous or calcified artery to reach its intended destination, and it may not closely conform to a bend. Stents with branch points, which can bolster both an artery and its communicating vessel, are still on the drawing board. What's more, once it is in place, a stent can cause scar tissue to grow or a clot to form. One likely solution is bioabsorbable stents—like the one Raj had implanted in his left main coronary artery—which are made from materials such as polylactic acid and coated with drugs to prevent scar tissue growth. Over time the material is absorbed (like an internal suture), leaving the expanded artery with its optimal diameter.

Cardiac surgery will also undergo major changes. Curiously enough, the biggest advance in this now commonly used mode of therapy will be that it will cease to be quite so common. This will be a big change from the heady days after the advent of open heart surgery in the 1950s, when major breakthroughs seemed to come every few years and many observers considered operations to be the ultimate step in the treatment of heart disease. Once surgeons were able to get their gloved hands on the cardiac muscle, the thinking was, all of its major problems could be solved.

Nevertheless, the trend will be away from the chest-cracking, rib-splitting surgery that requires eight to twelve weeks of recovery and leaves a ten-inch scar on the patient's chest. Minimally invasive techniques that are in early stages of clinical use involve making small incisions that allow instruments to be introduced between the ribs. Thoracoscopy, in which a miniature camera is inserted into the chest, allows the surgeon to visualize the heart on a screen or through a headset. Syndromes that once required open procedures such as a bypass grafting or valve replacement will increasingly be treated in the cath lab, with overnight stays taking the place of days in intensive care. Robotics will be perhaps the biggest development in surgery itself, with remotely operated devices being deployed everywhere from the battlefield to rural areas where economies cannot support a full surgical staff and facilities. Although it is impossible to say whether open heart surgery will disappear in the coming century, we will almost certainly see a radical drop in its frequency.

It is a brave new world, one in which the heart can be pictured as beating almost indefinitely, with physicians warding off once-fatal diseases with an arsenal of ever-less-invasive procedures that rely on the patient's body itself to provide the proper therapeutic tools. After all, if the heart is improvable—if we can run marathons after recovering from heart attacks, if we can make love after receiving a heart transplant—then might it not even be perfectible? Scientists at NASA have made nickel hydrogen batteries that can last for decades—could not the engines in our chests be altered to endure for that long as well? Is the future heart an eternal heart, as the Egyptians promised four thousand years ago?

And yet it is possible, as suggested in the fictional stories of Kirk and Raj, to envision a future in which heart disease will continue to be a major killer of both men and women. This has nothing to do with a slowdown in scientific progress, but rather with socioeconomic factors. Human frailty has been seen as an inextricable feature of the metaphorical heart from the Old Testament through Shakespeare and the Romantic poets and right up to our own time. In the future, all-too-human weaknesses threaten to impede the medical revolution in the treatment of the cardiac organ.

Statistics can be reassuring when we think about healing the heart; they can also be daunting. According to the American Heart

Association, ten years into the twenty-first century, one-fifth of the nation's high school students smoke—and more than six thousand new smokers join their ranks every day. Only just more than a third of high schoolers are physically active, and 15 percent of them are seriously overweight. Despite vigorous public-health education campaigns that counsel otherwise, many young people continue to eat foods that are fatty and salty and to spend their leisure hours parked in front of computer and television screens.

It is no better when they become adults. As just one example of this disturbing trend, the Institute of Medicine estimates that one in three adults in the United States has high blood pressure, which is the nation's second-leading cause of death. That is more than seventy million people, with another fifty million on the brink of developing the condition. Fewer than half of these people are adequately cared for, even though treatment involves such relatively straightforward measures as losing weight and cutting the amount of salt in the diet.

Perhaps this situation would be at least partially ameliorated if every one of these people had access to high-quality health care; then the latest technologies and medicines could be used to repair some of the damage being done by lifestyle. But it remains to be seen if the 2010 health care reform legislation will lead to affordable, adequate cardiac care for an overwhelming majority of Americans. The truth is that the boom in health care technology and innovative practices of the past half century has been accompanied by an explosion in cost that shows no sign of abating anytime soon. Before we can achieve a cardio-utopia in which medicine makes heart disease a rare occurrence instead of our primary killer, we must overcome obstacles that lie outside the reach of the laboratory and the operating theater. No matter how dazzling all of these high-tech medical therapies might be, they can't even collectively do as much as we can do to improve our hearts' health by taking simple steps to lower our blood pressure and cholesterol levels, exercise more, quit smoking, and reduce or even eliminate our consumption of trans fats and saturated fats.

In 2009 *JAMA: The Journal of the American Medical Association* published a research letter from a team of cardiologists who had recently performed an extraordinary procedure in Cairo. The group of four Americans and one Egyptian, specialists in cardiovascular imaging, had taken whole-body CT scans of twenty mummies at the Egyptian National Museum of Antiquities, looking for calcification in their arteries. The idea for the project had come about two years earlier, when Gregory S. Thomas, a cardiologist and professor at the University of California at Irvine, read a plaque describing the mummified pharaoh Menephtah that claimed he had died of atherosclerosis sometime around 1200 BC. Wondering whether such a diagnosis could be confirmed, Thomas decided to enlist some colleagues to bring the latest technology to bear on ancient remains.

Using a scanning system set up in a trailer beside the museum, Thomas and his team were able to detect hearts or major vessels in sixteen of the mummies. Of these, nine showed signs of arterial calcification in either their legs or their aortas, suggesting that they had suffered from some degree of coronary artery disease. The most ancient of these cadavers belonged to Lady Rai, a nursemaid to Queen Ahmose Nefertari, who died around 1500 BC. What intrigued the doctors about this discovery was that it seemed to fly in the face of the view that atherosclerosis is a very modern syndrome, a product of a fat-rich diet, smoking, a sedentary lifestyle, and the stress of urbanized living. One explanation for these results might be that all of those whose remains were scanned were associated with the pharaonic court and therefore would probably have had a diet rich in fatty meats such as duck and beef. The deceased also may have indulged in slothful lifestyles while others slaved away—literally—on their behalf.

Whatever this research tells us about atherosclerosis, that contemporary physicians examined mummified remains in a high-tech imaging machine speaks volumes about our ongoing fascination with the human heart. It tells us how much our understanding of the organ has advanced, yes, but also how little it has changed. After all, the only reason these modern doctors had a chance to examine the hearts of the Egyptian dead was that their predecessors, the physician-priests of the goddess Sekhmet, had purposely left the cardiac organs intact so those entombed could be judged in the hereafter. It was a ceremony that found an uncanny echo in that Cairo museum thousands of years later. Only this time, the sarcophagus into which the dead were placed was not a stone crypt, but rather a modern CT machine, with its slablike table and pure white ring. And those who judged them this time were not the gods, but doctors, mere mortals who could nevertheless draw upon the ceaseless toil and inspired visions of those who came before them to peer right into the core of each heart and see the truth it held.

Acknowledgments

Tom: I would like to thank David C. Sabiston, who taught me to appreciate the history of medicine, and Kanu Chatterjee, who taught me everything else. And of course, my wife, Jean, who has been with me through all of this.

Stephen: I would like to thank my wife, Caryl, for her patience, her intelligence, and her care.

And we would both like to thank Gary Beneze for his images of the heart, and Dr. Glenn Barnhart for his insights into heart surgery. We would also like to thank our agent, Henry Dunow, whose wisdom and encouragement helped turn a simple idea into a book. We would also like to thank Colin Dickerman and Gena Smith at Rodale, whose editing skills allowed us to write the book we'd always wanted.

Bibliography

Blair, Kirstie. *Victorian Poetry and the Culture of the Heart*. Oxford: Oxford University Press, 2006.

Boccaccio, Giovanni. *Decameron*. Translated by G.H. McWilliam. New York: Penguin, 2003.

Calvin, John. *Institutes of the Christian Religion*. Translated by Henry Beveridge. Peabody, MA: Hendrickson, 2008.

Conrad, Lawrence I., Michael Neve, Vivian Nutton, Roy Porter, and Andrew Wear. *The Western Medical Tradition, 800 BC to AD 1800*. New York: Cambridge University Press, 1995.

Elliot, Dyan. *Proving Women: Female Spirituality and Inquisitional Culture in the Later Middle Ages*. Princeton: Princeton University Press, 2004.

Erickson, Robert A. *The Language of the Heart, 1600-1750*. Philadelphia: University of Pennsylvania Press, 1997.

Fleming, P. R. *A Short History of Cardiology*. New York: Rodopi B.V. Editions, 1997.

Ford, John. *'Tis Pity She's a Whore and Other Plays*. New York: Oxford University Press, 2008.

Forssmann-Falck, Renate. "Werner Forssmann: A Pioneer of Cardiology." *The American Journal of Cardiology*. 79. (1997): 651-60.

Gregg, Pauline. *King Charles I*. Berkeley: University of California Press, 1981.

Gryllis, R. Glynn. *Trelawny*. London: Constable, 1950.

Harris, C.R.S. *The Heart and the Vascular System in Ancient Greek Medicine*. New York: Oxford University Press, 1973.

Harvey, William. *On the Motion of the Heart and Blood in Animals*. Translated by Robert Willis. New York: Prometheus, 1993.

Hawthorne, Nathaniel. *The Scarlet Letter and Other Writings.* Edited by Leland S. Person. New York: W.W. Norton, 2005.

Hoystad, Ole M. *A History of the Heart.* London: Reaktion, 2009.

Jager, Eric. *The Book of the Heart.* Chicago: University of Chicago Press, 2001.

Keynes, Geoffrey. *The Life of William Harvey.* Oxford: Clarendon Press, 1966.

King, Helen. *Greek and Roman Medicine.* Bristol: Bristol Classical Press, 2001

McNamara, Jo Ann Kay. *Sisters in Arms: Catholic Nuns through Two Millennia.* Cambridge: Harvard University Press, 1996.

Malloch, Archibald. *William Harvey.* New York: Paul B. Hoeber, 1929.

Miller, Jonathan. *The Body in Question.* New York: Horizon, 1986.

Pedley, John. *Sanctuaries and the Sacred in the Ancient Greek World.* New York: Cambridge University Press, 2005.

Poe, Edgar Alan. *Complete Tales & Poems of Edgar Alan Poe.* New York: Vintage, 1975.

Porter, Roy. *The Greatest Benefit to Mankind: A Medical History of Humanity.* New York: W.W. Norton, 1999.

Rosner, Fred. *Medicine in the Bible and the Talmud.* Jersey City: Ktav Publishing House, 1995.

Sawday, Jonathan. *The Body Emblazoned: Dissection and the Human Body in Renaissance Culture.* London: Routledge, 1996.

Shakespeare, William. *The Oxford Shakespeare: The Complete Works* 2nd ed. Edited by John Jowett, William Montgomery, Gary Taylor, and Stanley Wells. New York: Oxford University Press, 2005.

Shelley, Mary. *Frankenstein.* Edited by Maurice Hindle. New York: Penguin, 2003.

Slights, William W.E. *The Heart in the Age of Shakespeare.* New York: Cambridge University Press, 2008.

Trelawny, Edward John. *Letters of Edward John Trelawny*. Edited by H. Buxton Forman. New York: Oxford University Press, 1910.

Wickkiser, Bronwen. *Asklepios, Medicine, and the Politics of Healing in Fifth-Century Greece: Between Craft and Cult*. Baltimore: Johns Hopkins University Press, 2008.

Wood, Jane. *Passion and Pathology in Victorian Fiction*. New York: Oxford University Press, 2001.

Index